ORGANIC EXPERIMENTS

ORGANIC EXPERIMENTS

FIFTH EDITION

Walter W. Linstromberg
University of Nebraska at Omaha

Henry E. Baumgarten
University of Nebraska–Lincoln

D. C. HEATH AND COMPANY
Lexington, Massachusetts Toronto

Acknowledgment

Our thanks to Sadtler Research Laboratories for permission to use the infrared spectra labeled *Sadtler* and also to Varian Associates for the use of nmr spectra labeled *Varian*.

Published simultaneously in Canada.

Printed in the United States of America.

International Standard Book Number: 0-669-05524-7

Preface

The fifth edition of this manual, like its predecessors, is intended primarily for use in elementary organic courses having fifteen to thirty laboratory periods. The basic organization of the previous editions for the most part has been retained, with the first eight experiments introducing students to the important techniques of the organic chemistry laboratory. Following the introductory exercises are nine experiments (test tube, preparative, and unknown characterization) illustrating characteristic chemical properties of the various classes of compounds. These are arranged, more or less, in the order usually followed in introducing functional groups. The first seventeen experiments would easily suffice for a one-semester laboratory program, but could be supplemented or replaced by certain experiments which follow. The latter are largely synthesis experiments that include certain of the "name" reactions that have so richly contributed to the lore of organic chemistry. The last five experiments are multi-step syntheses or reaction sequences each of which may well require more than one laboratory period, and are intended for especially interested students and those of sufficient attainment.

In a continuing effort to make instructions unmistakably clear and to ensure good experimental results, some of the experiments have been revised both narratively and experimentally. The manual contains no experiments that the authors consider to be more hazardous than those usually carried out in beginning student laboratories; however, in accord with current trends in every field of endeavor, the number and scope of caution alerts that warn the student and instructor of potential hazards, however remote, have been increased. Furthermore, in keeping with good laboratory practice, which assigns first priority to safe procedures and minimum or nonexposure to toxic compounds, especially to those now classified as OSHA Category I chemicals, the authors have limited experimental procedures to those that avoid the use of *carbon tetrachloride*, *chloroform*, and *benzene*.

The fifth edition includes seven new experiments to give the instructor still greater flexibility in determining a laboratory program best suited to the students. Experiment 18 illustrates the biosynthesis of ethanol and acetic acid from apple cider. It is a two-part experiment that not only illustrates an historical and commercially important synthesis, but also affords an opportunity for the student to practice certain techniques and review reactions previously learned. Experiment 34 is a rather lengthy sequence embodying several steps that finally culminates in the classical Diels-Alder reaction. This experiment requires the preparation and use of 2,3-dimethyl-1,3-butadiene within one laboratory period. The diene, in turn, is prepared from pinacol (Experiment 33)—a challenging synthesis in itself. A diversion of the sequence appears in

Experiment 35 in which pinacol is rearranged to pinacolone and provides a route to pivalic acid via the haloform reaction. Experiment 35 illustrates the value of the haloform reaction as a preparative method, in addition to its more familiar use as a diagnostic test.

As in previous editions, special attention has been given to the economy of materials. With few exceptions reagents, chemicals, and unknowns of relatively low cost have been selected, with quantities used kept as small as practical. Wherever possible, student preparations are used as unknowns in identification exercises. In sequential experiments examples have been chosen such that all intermediates are prepared by the student or are available and relatively inexpensive. An exception is found in Experiment 31, which begins with *tert*-butylbenzene, a rather expensive material. This experiment may be modified to begin with commercially available bromobenzene, which will lead to the preparation of unsubstituted benzoic acid and thus avoid the use of benzene.

Time schedules are included for all experiments. These schedules should be considered only as rough guides, since efficiency in the laboratory varies with the individual student, as every laboratory instructor is fully aware. The schedules are estimates based on experience with mixed groups of majors and nonmajors under reasonably close supervision. When preparations require longer periods than allowed for in a laboratory session, a safe point at which the experiment may be interrupted is indicated.

New questions and exercises have been added to nearly all experiments. These vary in difficulty and are intended to explore various facets of the chemistry, technique, mechanism, etc. of the procedure under study; some examine alternative methods for carrying out the experiment and suggest reasons for carrying it out. A number of study problems are included in the fifth edition that use ir and nmr spectra as useful factual information required for their solution. These exercises have an asterisk and are included as optional exercises for especially interested students.

The authors appreciate the wide acceptance of the previous editions and are grateful for the many helpful comments and suggestions that have been made by users. A number of changes in experimental procedures in the fifth edition as well as in previous editions are based on suggestions by users, and to these persons the authors are indeed indebted.

As before, each user is invited to consider this collection of experiments as his or her manual, to feel free to improvise and improve upon it, and to send these findings to the authors, because every successful organic chemistry laboratory manual represents the contributions of many teachers, past and present.

WALTER W. LINSTROMBERG
HENRY E. BAUMGARTEN

Contents

Appendix

ORGANIC EXPERIMENTS

General Instructions

The purpose of laboratory work in organic chemistry is twofold: (1) to illustrate the reactions and general principles discussed in lectures and your textbook and (2) to acquaint you with those laboratory practices and techniques which make up what has been called the **art of organic chemistry**. It soon will become apparent to you that reactions in organic chemistry are far different from those previously encountered in your general, or inorganic, chemistry course. Organic reactions frequently require long periods of time (even at elevated temperatures) for completion, and complicating side reactions that lead to a mixture of products are quite common. Considerable manipulation and treatment are required for the isolation of most pure products, and final yields may be much less than those theoretically possible. In general, the acceptability of an organic reaction is governed not only by the quantity of product obtained from the reaction but also by the quality (purity) of the product. A high yield of an impure substance is no better and, in some cases, is far worse than a low yield of a pure product.

For all the seemingly unkempt appearance of most organic chemistry laboratories, no branch of chemistry requires a higher development of laboratory techniques or a greater refinement of laboratory practices than organic chemistry. The organic chemist has developed a number of laboratory operations and many simple pieces of laboratory equipment that will be unfamiliar to you but which have been found to be indispensable in the practice of this science. Learning proper procedures and techniques in the organic laboratory will aid you in other science courses that follow, especially those in the medical, pharmaceutical, and biological sciences.

Follow directions carefully, ask questions whenever the procedure is not clear to you, and exercise care in the purification of any products made. These are the simple directions that will lead to a proper achievement of the purpose of the laboratory. You are expected to work alone unless otherwise instructed. Address all questions concerning the experiments to your laboratory instructor, *not to your neighbor*. Results, when obtained, or observations, when made, should be recorded immediately and directly on the report forms provided. All experiments with the exception of multi-step preparations should be completed within the prescribed laboratory period. Clean all apparatus at the *end* of each experiment rather than before. It is always more easily cleaned after recent use; moreover, it is better to begin an experiment with clean and *dry* apparatus.

Ground-Glass Equipment

Cork- and rubber stopper-type glassware once commonly used in beginning organic laboratories has been replaced to a large extent by tapered ground-glass, jointed glassware. This equipment, usually referred to as standard-taper ($) glassware, has a number of advantages. All pieces of equipment with ground-glass joints of one size are interchangeable. The need for boring stoppers is eliminated and apparatus can be assembled in far less time. Assemblies illustrated in this manual show standard-taper ware. However, the use of standard-taper equipment requires the exercise of special techniques. Such ware, when connected, makes a rigid joint, and an assembly must be properly supported to prevent any undue strain on the parts. Moreover, standard-taper glassware must be assembled in a definite order — from the bottom up. A lower section should not be expected to hold fast to an upper section simply because it fits snugly. If the lower section is not supported, it may slip off and break. A simple expedient for the prevention of accidental breakage is the use of rubber bands to hold sections together while assembling. Finally, all ground-glass surfaces must be lightly greased to make an airtight seal and to prevent these surfaces from "freezing."

Laboratory Reports

Forms for reporting results appear at the end of each experiment. Each report form provides space for all pertinent information needed to describe the experiment and to evaluate results. Reaction equations, quantities required (expressed in grams and moles), physical constants, percentage yields, results of tests, etc., are all provided for. Data should be recorded as the experiment is performed — not hours after it is completed. The report forms are detachable and provide a convenient and uniform record of work done. However, should the instructor prefer that students keep a laboratory notebook, the report forms may be detached and secured with a strip of magic tape to the right-hand pages of a hardcover, bound notebook, reserving the left-hand pages for observations, comments regarding the experiment, and answers to the questions that follow each report. All reporting should be done in ink.

Certain of the synthesis experiments require that you submit the purified product to your instructor. Inasmuch as you have spent several hours preparing it, the fruit of your labor deserves to be placed in a clean, tightly-stoppered vial with all pertinent information on an affixed, legible label. The label should include the name of the compound and structure (if space permits), the gross and tare weight of the vial, the net weight of the contents, observed bp or mp, name of student, and date. A sample label is shown below.

Acetanilide, (benzene ring)—N(H)—C(=O)—CH_3

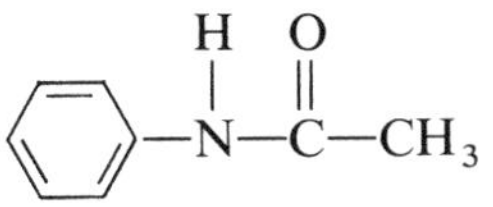

Gross Wt. 37.2 g	mp 114°
Tare Wt. 24.2 g	
Net Wt. 13.0 g	
John Smith	Dec. 16, 198__

Experiment 1

Time: $2\frac{1}{2}$–3 hours

Melting Points

The identification of a new organic compound is often a laborious and exacting task. A known compound, on the other hand, usually may be identified (or characterized) by the determination of one or more readily measurable **physical properties** (melting point, boiling point, refractive index, infrared, ultraviolet, or nuclear magnetic resonance spectrum, etc.) in conjunction with the examination of a few **chemical properties.** For solid substances one of the most useful physical properties is the melting point. The melting point of a substance is defined by the physical chemist as the temperature at which the solid and liquid phases are in equilibrium. The determination of such melting points as a routine identification procedure is difficult, especially with a limited amount of material. Fortunately, such melting points are unnecessary in the work of the organic chemist. The organic chemist utilizes instead a **capillary melting** point. The capillary melting point is a *range* of temperature over which a minute amount of the solid in a thin-walled capillary tube first visibly softens and finally completely liquifies. For the sake of simplicity, such melting point ranges are called **melting points** (abbreviated mp) and have been utilized by organic chemists for a century or more. These melting points are just as meaningful to him as the melting points of the physical chemist, and all of the recorded melting points in handbooks, journals, and other organic chemical literature, unless otherwise specified, are of this type.

The melting point range of a solid organic compound, if very nearly pure, will be small (0.5–1.0°), and the substance is said to melt **sharply.** The presence of impurities, even in minute amounts, usually depresses, or lowers, the melting point and widens the range. Thus, the melting point not only is useful as a means of identification but also as a criterion of purity. Depression of the melting point of a pure compound by an impurity also is useful as an aid in identification. Consider, for example, a substance X thought to be identical with the compound Y because X and Y have the same melting point. If we mix a small amount of X and Y together and measure the melting point of the mixture, we determine what is called the **mixture melting point** of X with Y. If the mixture melting point is the same as the melting point of Y, X and Y *probably* are identical. (Exceptions to this generalization are relatively uncommon but are known.) However, if the mixture melting point is depressed below the melting point of Y (and the range broadened), then X acts as an impurity in Y and the two cannot be identical.

In this experiment you will determine (a) the melting points, separately, of a pair of pure organic compounds which melt at or near the same temperature, (b) the mixture melting points of the same pair, and (c) the melting point of an unknown compound.

Procedure

A. MELTING POINT APPARATUS

A laboratory melting-point apparatus should meet several design objectives. First there must be a convenient source of heat, usually a gas flame or an electrical resistance heating element. Second, the temperature rise in the vicinity of the sample must be controllable. With a gas flame or an electrical element this may be accomplished by using a reasonably large volume of some thermally stable liquid, usually mineral oil or silicone oil (Fig. 1.1) so that most of the heat being applied is expended in heating the oil rather than the very small sample. Electrical devices may employ a large block of metal as a heat sink or a carefully designed low-wattage heating element and a smaller block (Fig. 1.3). Third, the temperature in a reasonably large zone in the vicinity of the sample must be as nearly uniform as possible so that both the sample and the temperature measuring device are at the same temperature. This is usually accomplished by convection currents in the oil (Fig. 1.1), or, in the case of a sophisticated electrical device, by careful engineering and, possibly, insulation of the heat sink or block (Fig. 1.3). Finally, some means of measuring the temperature must be provided. Although this can be as elaborate as a relatively expensive thermistor device, in all of the apparatus set-ups described here a laboratory thermometer is used. The thermometer must be oriented so that its bulb is as close to the sample as possible (i.e., in the zone of uniform temperature). In careful work the thermometer will be "calibrated" *in the apparatus* by determining the melting points of a series of *pure* compounds whose melting points cover the range of the thermometer.

Both of the apparatus set-ups described here will give satisfactory melting points. They are presented in order of increasing complexity and cost. The electrically heated device described in procedure (A.2) is only one of several well-designed commercial melting-point devices currently available. It is described because students in the authors' laboratories have indicated a preference for it over other devices made available to them.

(A.1) THIELE TUBE

If Thiele melting-point tubes are provided, assemble a melting-point apparatus like that illustrated in Figure 1.1.

Support a Thiele tube on a ring stand, using a buret clamp, and fill the tube to the top of the upper arm with clear mineral oil. Carefully bore a hole in a cork large enough to accommodate a 260°C thermometer. With a sharp knife or razor blade cut out a one-sixth section of the cork and slide it over the thermometer in such a way that the calibrations are visible through the cut-out opening. Fasten the cork and thermometer in a buret clamp centered about six inches above the Thiele tube. Lower the thermometer into the Thiele tube by sliding it through the cork until it is about one inch from the bottom of the

FIGURE 1.1 The Thiele melting point apparatus.

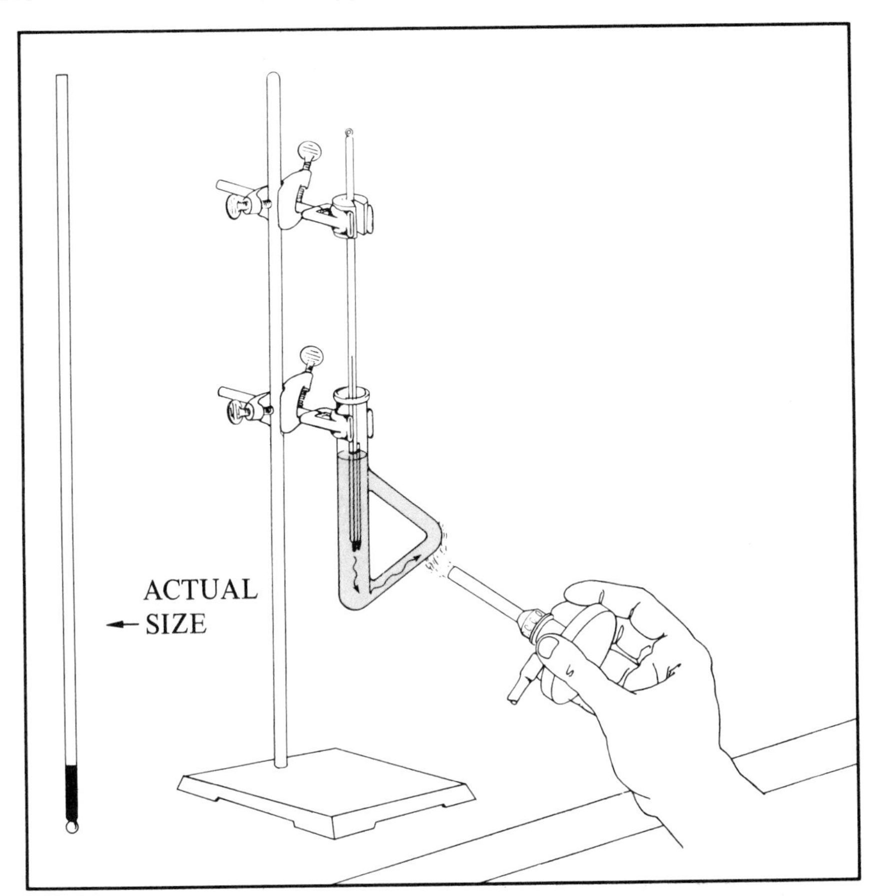

straight portion of the tube. The buret clamp, when closed, will compress the cut cork and grip the thermometer quite firmly.

(A.2) COMMERCIAL ELECTRICALLY HEATED APPARATUS

The Mel-Temp® capillary melting-point apparatus (Note 1)[1] is illustrated in Figure 1.3. This device has a sealed electrical heating element, which provides a zone of uniform temperature by means of a high thermal conductivity block, heats each capillary on three sides, and accepts up to three capillary melting-point tubes at one time. The melting-point tube is illuminated by a built-in light source and is observed through a six-power lens system. To

[1] This reference is to notes usually found at the end of the procedure.

use the apparatus, a melting-point tube is inserted in one of the capillary wells, the voltage of the heating element is set to obtain the desired heating rate at the anticipated melting point using the voltage control and temperature vs. time or voltage curves (Fig. 1.2), and the melting point is observed through the viewing lens. CAUTION! *Be sure to turn off the apparatus after each use to avoid serious damage to the heating element.*

FIGURE 1.2 Heating curves of Mel-Temp at different fixed voltages.

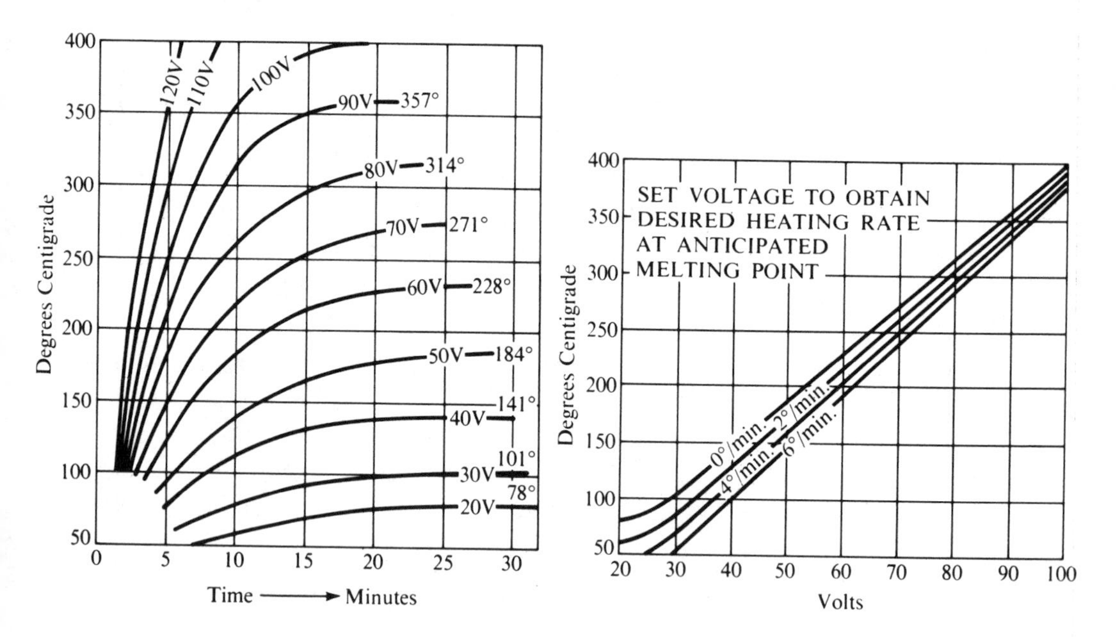

B. MELTING POINTS OF PURE COMPOUNDS

Obtain six melting-point capillaries from the storeroom or your instructor. Seal one end of each capillary by slowly rotating the end in the edge of a hot flame until a small bead appears (see Fig. 1.1). When cool, these capillaries are ready for filling. Have your instructor check your melting-point apparatus and your capillary tubes before proceeding further.

Place a small sample of any one of the 21 compounds listed in Table 1.1 on a piece of clean paper or clay plate. Crush the material to a fine powder with the spatula and scrape it into a small mound. Fill one of the melting-point tubes by pushing the open end into the mound of powder, using the

FIGURE 1.3 Commercial electrically heated capillary melting point apparatus.

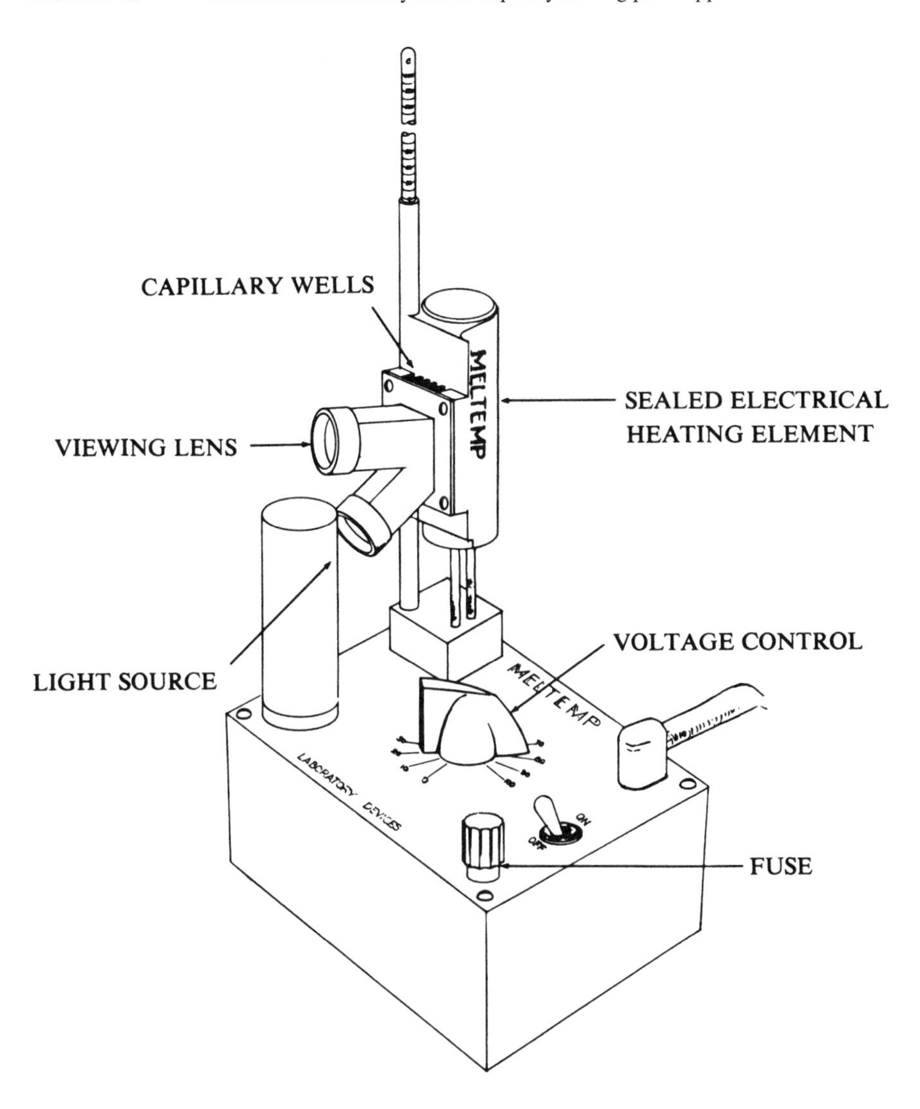

spatula as a backstop. When a small plug of powder has collected in the opening of the capillary tube, work the material down to the sealed end by scratching the capillary with a file while holding it lightly at the top. Repeat this process until a column of powder about 3–5 mm in height has collected in the capillary tube. Tamp the powder compactly in the capillary by dropping it on the desk several times through a 2-foot length of glass tubing.

TABLE 1.1
MELTING POINTS OF SOME COMMON ORGANIC COMPOUNDS

Compound	mp, °C
Pimelic Acid	103–105
Catechol	104–106
Azelaic Acid	105–106
Resorcinol	109–110
Acetanilide	113–114
dl-Mandelic Acid	117–118
Succinic Anhydride	118–120
Benzoic Acid	121–122
2-Naphthol	121–122
Urea	132–133
trans-Cinnamic Acid	132–133
Maleic Acid	134–136
Malonic Acid	135–137
Benzoin	136–137
Anthranilic Acid	145–147
Cholesterol	148–150
Adipic Acid	152–153
Citric Acid	153–155
Salicylic Acid	156–158
Benzanilide	160–161
Itaconic Acid	163–165

If you are using an apparatus like that in Figure 1.1 follow the procedure given in the next three paragraphs. If you are using the electrically heated device shown in Figure 1.3, *read* these paragraphs and modify the procedure as directed in the final paragraph of this section.

Attach the capillary tube to the thermometer with a small rubber band sliced from the end of a piece of rubber tubing. Position the capillary on the thermometer so that the filled portion is just opposite the mercury bulb. Replace the thermometer in the bath as before. Make sure that the small rubber band is well above the surface of the oil or the rubber will soften and disintegrate during the determination.

Heat the oil with a small flame and allow the temperature to rise fairly rapidly to within 15–20° below the melting point of the compound; then the burner should be adjusted so that the temperature rises no more rapidly than 2–3° per minute during the actual determination of the melting point. Watch the sample closely and check the temperature reading frequently. Write down the temperature at which the first visible softening of the sample is noted. Continue heating at 2–3° per minute and note the temperature at which all of the material has turned to liquid. The two values determined define the melting range or the melting point. Record your results.

Repeat the above process using a different compound with a melting point at or near that of the first. Discard melting tubes after one use and do not attempt to redetermine a melting point on a sample previously melted.

If you are using the electrically heated apparatus of Figure 1.3, modify the above directions as follows. Rather than attach the capillary melting-point tube to the thermometer, insert the melting-point tube in one of the three capillary wells. Rotate the voltage control knob to "0" and turn on the apparatus. Make certain that the sample in the melting-point tube is visible in the viewing lens. Adjust the voltage control knob to a setting that will give a reasonably rapid rise (about 5 minutes) of temperature to within about 15–20° below the anticipated melting point. (To determine the voltage setting to be employed, use the family of curves in Figure 1.2 in which temperature is plotted against *time*.) When this temperature has been reached, quickly lower the voltage setting to a value that will give a heating rate of 2–4° rise per minute during the actual determination of the melting point. (Obtain the voltage setting to be used from the family of curves in which temperature is plotted against *voltage*.) You should be able to get satisfactory results by using an initial voltage setting of 60 volts, then at a temperature of 15–20° below the anticipated melting point quickly lowering the voltage to 35 volts if your sample should melt between 100 and 120°, to 40 volts if it should melt between 120 and 140°, or to 45 volts if it should melt between 140 and 160°. The observation and recording of the melting point is carried out exactly as described above for melting points determined in an oil bath. After you have acquired confidence in your ability to use the instrument and interpret the heating curves, you may use it to determine as many as three melting points at once of different substances, to compare the melting points of different samples of the same substance, or to make mixture melting-point determinations [as described below in procedure (C)]. Be sure to turn the voltage control to "0" and turn off the apparatus as soon as you have finished.

C. MIXTURE MELTING POINTS

Take approximately equal amounts of two compounds selected from Table 1.1 and mix thoroughly on a clay plate or a clean piece of paper. Better still, grind them into an intimate mixture with a mortar and pestle. Determine the melting point of the mixture and note carefully the effect one compound has upon the melting point of the other.

D. DETERMINATION OF AN UNKNOWN

Obtain an unknown compound (it will be one listed in Table 1.1) from your instructor, record its number, and determine its melting point in the following way. Fill two melting-point tubes with the unknown. Determine an approximate melting point using one tube and a rate of heating of about 15–20° rise per minute. If you are using the apparatus shown in Figure 1.3 for this experiment, a voltage setting of 65 volts should suffice. As soon as the sample melts, remove the burner and allow the bath to cool to 20° or more below the approximate melting point. Then quickly attach the second capillary tube to the thermometer and determine an accurate melting point using the approximate melting point as a guide in the determination. This procedure should be followed with all samples for which the melting point is not known in advance.

Identify your unknown from its melting point and from its mixture melting points with an identical compound from the side shelf (Table 1.1).

Note 1. The Mel-Temp capillary melting-point apparatus is available from Laboratory Devices, P.O. Box 68, Cambridge, Massachusetts 02139. It is advisable to order a number of extra light bulbs with the apparatus; the life of these bulbs is unpredictable. Note also that much shorter capillaries are required for this apparatus than for most high grade oil baths. Depending upon the size of capillary stocked, it may be possible to cut the capillaries in half and use each half to make a melting-point tube. The capillary tube should project out of the well just far enough to facilitate removal of used capillaries.

TEXT REFERENCE

Report: 1

Chapter Pages

Section Desk NAME ______

MELTING POINTS

MELTING POINTS OF PURE COMPOUNDS

Compound name	mp °C (Experimental)	mp °C (Literature)
(a)		
(b)		

MIXTURE MELTING POINTS

Melting point of a mixture of (a) and (b) ______°C

IDENTIFICATION OF AN UNKNOWN

Number of unknown ______

mp of unknown ______°C

mp of unknown when mixed with ______°C
(name of compound)

mp of unknown when mixed with ______°C
(name of compound)

mp of unknown when mixed with ______°C
(name of compound)

Unknown compound is ______

Questions and Exercises

1. Give two reasons why a sample should be finely powdered and tightly packed when a capillary melting-point determination is to be made.
2. What effect would the following nonorganic substances have upon the melting point of an organic compound: (a) sand; (b) water; (c) salt?
3. What effect would using too large a sample have upon the melting point of a compound?
4. What effect would too rapid heating have upon the apparent melting point of a compound?
5. Some substances, particularly high-melting compounds, give broad, indefinite melting points when the determination is carried out in the usual manner but give sharp, well-defined melting points when the sample is not placed in the bath (or block) until the temperature is about 20° below the melting point of the substance. Suggest an explanation for this behavior.
6. What properties other than those mentioned in the introduction are often valuable as aids in the identification of an organic compound?
7. Name five different natural products of organic origin, which lack sharp melting points but soften over a wide temperature range. Are these pure organic compounds or mixtures?
8. One reason why you were instructed not to attempt to redetermine a melting point on a sample previously melted is that the once-melted sample may give either a substantially lower *or higher* melting point. What sorts of physical or chemical changes could be taking place during melting that would explain this behavior?
9. Phenol, C_6H_6O, is a colorless, crystalline solid with a melting point of 40–41°C, yet the phenol found on a stockroom shelf often appears in the form of a pink slurry rather than as a solid. Offer an explanation for this.
10. How would you determine the melting point of cyclohexane (6.5°) and 1,3-dibromobenzene (−7°)?
11. A German chemist refers to glacial acetic acid as *eisessig* (ice vinegar). Why?
12. A very small sample of a low-melting material disappeared from the capillary before the temperature ever reached its reported melting point. Explain.

Experiment 2

Time: 3–3½ hours

Recrystallization. Filtration

Organic compounds usually are more soluble in hot solvents than in cold. An impure solid organic compound, when dissolved in the proper amount of an appropriate solvent at an elevated temperature will reprecipitate when the solution is cooled. If the hot solution is filtered, before being allowed to cool, dirt, lint, or other insoluble impurities will be removed, and the crystals that deposit in the cooled solution usually will be more nearly pure than the starting material. The crystals may be removed from the filtrate (mother liquor) by filtration. The soluble impurities and a small amount of the desired substance will stay in solution. This process is called **recrystallization** and is one frequently used for the purification of solid organic compounds. The success of the process depends on the fact that soluble impurities usually are present in smaller amounts than the desired compound so that the cooled solution, although saturated with respect to the desired product, may not be saturated with the impurity. This being the case, the latter will not precipitate from the solution. Sometimes a solution does become saturated with respect to an impurity during the cooling process. In this event, the impurity deposits along with the desired product and the recrystallization will have to be repeated, perhaps several times. In each recrystallization a small amount of the desired compound remains behind in the mother liquor. Such losses are unavoidable if a pure product is to be obtained.

A good solvent for recrystallization has the following properties: (1) it dissolves a reasonable amount of the organic compound at high temperatures (usually the boiling point of the solvent) and very little at low temperatures, (2) it dissolves impurities readily at low temperatures or does not dissolve them at all, (3) it does not react with the substance being purified, and (4) it is readily removed from the purified product. The last requirement generally means that the solvent must have a relatively low boiling point so that it will evaporate readily. Sometimes a solvent with all of these properties cannot be found. In such instances the solvent that most nearly approaches the ideal is selected.

The process of recrystallization often may be aided, especially when the impurities are colored, by selectively adsorbing contaminants on activated charcoal. A small amount of charcoal is added to the hot solution just before the filtration step. However, charcoal will adsorb not only the impurities but also a certain amount of the desired product. It is advisable, therefore, to use a minimum amount of charcoal. A small micro-spatula measure usually will be ample for the experiments in this manual.

Inasmuch as many organic compounds are quite bulky and filter slowly, gravity filtration is seldom used in organic chemistry for the collection of precipitates. Suction filtration employing specially designed funnels usually is used to effect rapid filtration and drying of precipitates. For large amounts of material the Büchner funnel (Figure 2.1) is used. Small amounts of material may be filtered more conveniently by means of a Hirsch funnel (Figure 2.2).

FIGURE 2.1 Büchner funnel and filter flask for suction filtration.

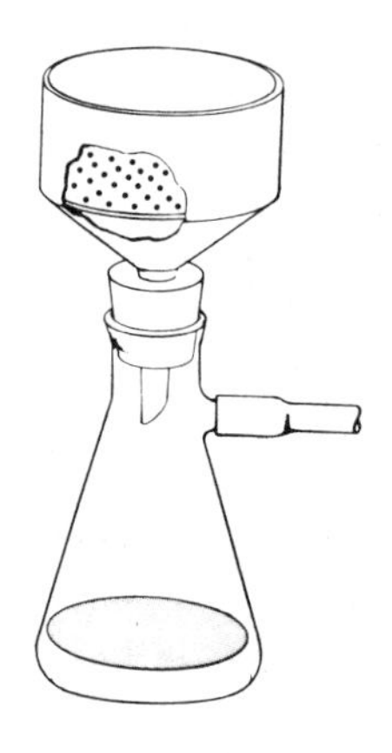

FIGURE 2.2 Hirsch funnel and filter flask for suction filtration.

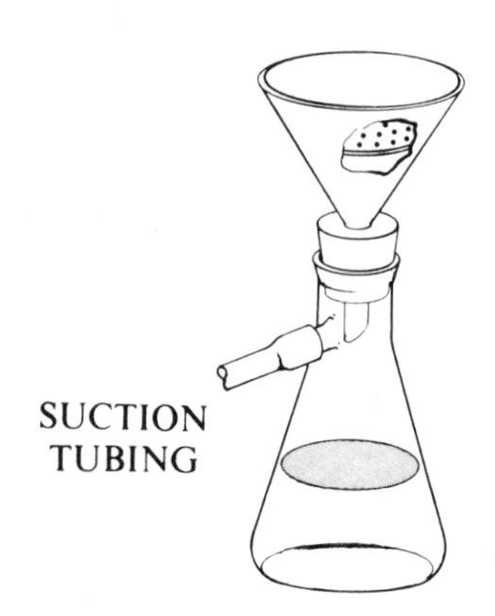

Procedure

Frequently it is necessary to heat organic compounds in a solvent for long periods of time without boiling away the solvent. This is accomplished by condensing the hot solvent vapor in a condenser and returning it as liquid to the boiling flask. A continuous boiling and back-flow of condensate in this manner is called **refluxing.**

Figure 2.3 shows an apparatus with a condenser set for refluxing. Assemble a reflux apparatus as illustrated in Figure 2.3 using a 50-ml round-bottomed flask and a small condenser. The assembly should be supported on a wire gauze about 2 inches above your burner. (Note positions of clamps on flask *and* condenser.) Make hose connections to the condenser.

Weigh out a 2.0-g sample of impure acetanilide. Save just enough for a melting point and add the rest to the 50-ml flask of your semi-macro reflux apparatus. It will be necessary to remove the *reflux condenser* temporarily to add the acetanilide. Be sure to reassemble the apparatus properly. Add 30 ml of water through the top of the condenser. Attach the lower end of the water jacket of the condenser to a water outlet and the upper end to the sink or drain, using a length of rubber tubing in both cases. A condenser thus is kept filled with cold water simply by allowing it to overflow. (See Fig. 2.3.) Circulate water through the condenser in a slow but steady stream. Bring the water in the small flask to a gentle boil by heating with the burner. Adjust the burner so that the water refluxes in a steady drip from the bottom of the condenser.

FIGURE 2.3 Apparatus assembly for refluxing liquids using standard-taper (Ⓣ) glassware.

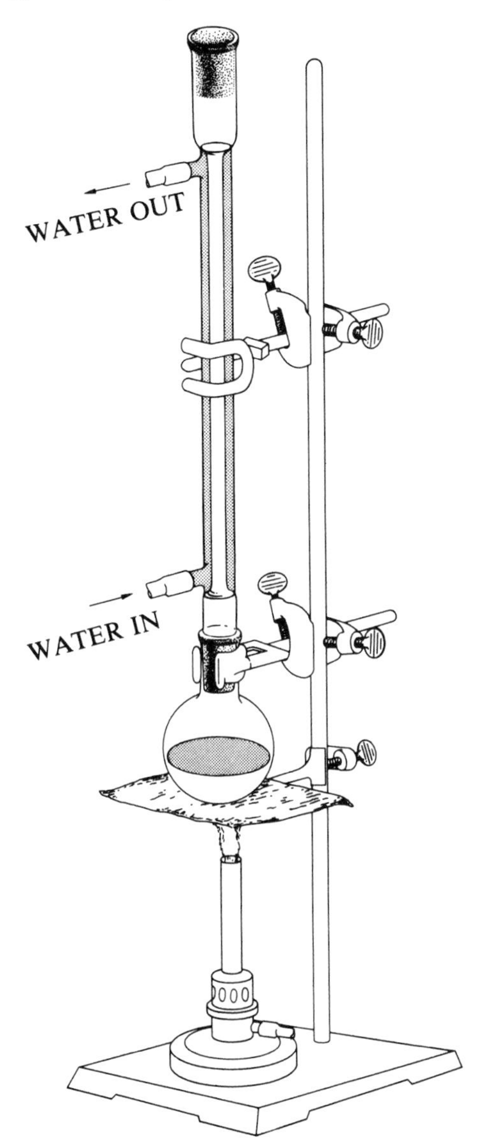

Continue to heat the mixture until no more solid appears to dissolve. Remove the burner and *allow the flask to cool a moment.* Then remove the condenser and cautiously add a *very* small amount (match head size on a micro spatula) of decolorizing charcoal to the hot solution. Replace the condenser and boil the mixture five minutes longer.

While the mixture is being heated with charcoal, fit your filter flask with the Hirsch funnel using a tight-fitting rubber stopper. Connect the filter flask to the aspirator with a length of suction tubing. Place a small filter paper disk in the funnel and turn on the water through the aspirator (Note 1). Preheat the suction apparatus by pouring 50 ml of boiling water through it. Discard the hot water which collects in the filter flask. Turn off the burner and remove the reflux condenser. Using the flask clamp as a handle, filter the **hot** acetanilide solution without delay. Transfer the filtrate (which probably will already contain some crystals) to a 50-ml beaker and cool by placing the beaker in a pan of ice. Discard the filter disk and the collected impurities, rinse the funnel clean and refit with a new paper filter.

When the filtrate is cold, collect the crystals by suction filtration. Moisten the crystals in the funnel with a little *ice-cold* water and dry them as much as possible with suction. Transfer the crystals to a clean piece of filter paper supported on a watch glass. Cover them with a piece of white paper and store in your desk for drying until the next laboratory period.

Weigh your dried product and determine its melting point as well as that of the crude material. Calculate the percent recovery of pure material. Turn in your purified acetanilide with your written report.

Note 1. Sudden drops in water pressure may cause water to be drawn into the vessel being evacuated. Although most aspirators are equipped with check valves to seal the system in the event of such a pressure drop, these valves often fail to function. Therefore, it may be advisable to insert a trap (Figure 2.4) in the line between the aspirator and the filter flask (or other evacuated vessel). This trap is constructed from a 250- or 500-ml suction flask, a stop-cock, a length of glass tubing with a right-angle bend, and a two hole stopper. In the interest of economy the stop-cock may be omitted; however, it provides easy control over the amount of suction or vacuum in the system and facilitates release of the vacuum at the end of an experiment.

FIGURE 2.4 Trap for water aspirator vacuum systems.

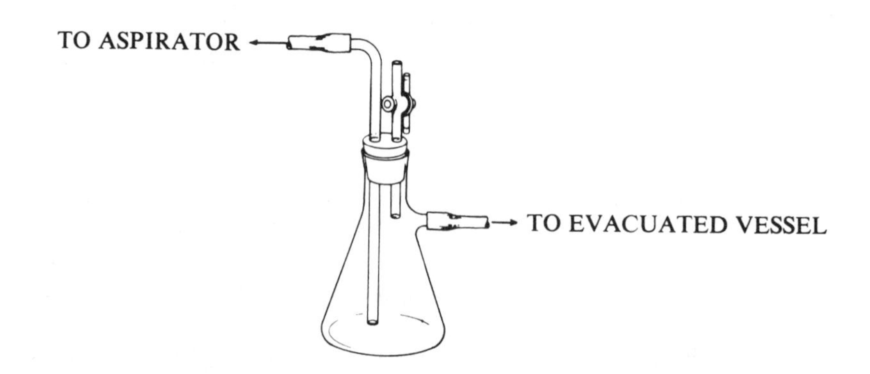

TEXT REFERENCE

Report: 2

Chapter Pages

Section Desk NAME

CRYSTALLIZATION

Amount of pure acetanilide recovered ________g

Percentage recovery ________%

Melting Points (°C)		
(impure)	(recrystallized)	(literature mp)

Questions and Exercises

1. A rule frequently followed by early organic chemists was stated in scholarly language as *simila similibus solvunter* or "like dissolves like." Restated, this means that a polar solvent will dissolve a polar substance, and a nonpolar solvent will dissolve a nonpolar substance. With this as a guide, what solvent would you use to remove a grease spot from a woolen skirt or from a pair of flannel trousers? What would you use to remove pancake syrup from these garments?
2. Cite several reasons why suction filtration is to be preferred to gravity filtration.
3. Why is a minimum amount of *cold* solvent used for washing precipitates already collected on a filter?
4. A compound is recrystallized from methanol, dried, and examined for purity by determination of its melting point. In a capillary the compound appears to melt at 115° with vigorous evolution of a gas; then the compound solidifies in the hot bath and does not melt again until 165°, at which temperature it again appears to melt sharply. Suggest an explanation for these observations.
5. Why is a mixture of two solvents sometimes necessary for the recrystallization of an organic compound?
6. The formation of crystals from a supersaturated solution, if not spontaneous, may be initiated by simply scratching the inner wall of the vessel containing the solution. Suggest a reason for this.
7. Frequently, a student will attempt to recrystallize a sample from a hot solution only to find that oil globules have formed rather than a homogeneous solution. What has been done wrong and what corrective measures should be taken?
8. Crystals often are separated from the mother liquor by gravity filtration through a filter paper that has been repeatedly folded or "fluted." What is the purpose of using a fluted filter paper?
9. Why should a filtration as described in Question 8 not be used for filtering a hot solution?
10. A crystalline compound is soluble in solvent A and in water. Solvent A is miscible with solvent B, but the material in solution is not soluble in B. The compound is now in solution in A. Suggest a plan for recovering the dissolved material.

Experiment 3

Time: 2½ hours

Distillation

The space above the surface of a liquid always contains some of the substance in the vapor (gaseous) state. If the container is an open one such as a beaker, the molecules of the vapor escape to the atmosphere to be replaced by other molecules escaping from the liquid. When this process takes place below the boiling point of the liquid it is called **evaporation.** If the container is a closed system such as a corked bottle, an equilibrium is reached in which the air space above the liquid is saturated with molecules of the vapor. The pressure exerted by the vapor in equilibrium with the liquid is called the **vapor pressure** of the liquid. The vapor pressure of a liquid is constant at a given temperature and is not affected by the total pressure. It is a measure of the tendency of the liquid to pass into the vapor state. It increases as the temperature is raised and decreases as the temperature is lowered. The **boiling point** of a liquid is the temperature at which the vapor pressure becomes equal to atmospheric (total) pressure. Thus, as the atmospheric pressure, or total pressure, is raised the boiling point also is raised; as atmospheric, or total pressure, is lowered the boiling point is lowered. If the temperature of a liquid is raised until its vapor pressure becomes equal to that of the atmosphere, i.e., to the boiling point, and maintained at that temperature, the liquid will pass entirely into the vapor state. This process is known as **distillation.** If the vapor of the liquid is conducted to another container which is cooled below the boiling point, it will **condense** (return to the liquid phase). When the chemist speaks of distillation he generally means the combined process of distillation followed by condensation.

Pure organic compounds distill over a very narrow range of temperatures called by the organic chemist the **boiling point** (bp). Any liquid with a wide-boiling or distilling range is impure, although a liquid with a very narrow, constant, boiling point range is not necessarily a pure one. The reason for this is that various compounds are affected in different ways by impurities. Some show no change in boiling point, others display elevated boiling points, and still others show depressed boiling points. For this reason the boiling point of a substance, while valuable, is not as good a criterion of purity as is the melting point.

Volatile liquids can be separated from nonvolatile substances by distillation, and, frequently, mixtures of volatile liquids can be separated into the component parts by **fractional distillation.** Occasionally two or more liquids form a constant-boiling mixture or **azeotrope,** which boils at a constant temperature and is made up of a fixed composition of the components. For example, pure ethyl alcohol boils at 78.4° and pure water at 100°, but a mixture of 95% ethyl alcohol and 5% water boils at 78.1°, and this mixture cannot be separated by distillation into pure water and pure ethyl alcohol. Other mixtures having different percentage compositions of ethyl alcohol and water can be separated by distillation into one of the components and the azeotrope.

During distillation a particular portion of the liquid may become momentarily heated above the boiling point, a large amount of vapor may form suddenly, and the contents of the flask may "bump," possibly carrying liquid up through the side arm of the distilling flask and causing the apparatus to be jolted severely. Bumping may be prevented to a large extent by adding a few pieces of porous clay plate or **"Boileezers".**[1] These mineral substances contain considerable air in their porous structures which act as a source of tiny bubbles, first of air, later of the volatile components. The vapors of the liquid form around these bubbles with little local superheating.

Procedure

Arrange an apparatus for distillation like that illustrated in Figure 3.1.

The thermometer should be positioned in the upper connecting adapter far enough to allow the tip of the mercury bulb to extend about 10 mm *below* the side arm opening (see Fig. 3.1). Assemble the apparatus after all glass joints have been lightly lubricated by supporting the distilling flask on a wire gauze and clamping it at the neck. Support the condenser with a second clamp and position it so that it is at the proper height and angle to join the connecting adapter. Make any minor adjustments in clamps and ring stands necessary to join condenser to distilling flask in a strain-free assembly. Attach the take-off adapter to the end of the condenser (a rubber band looped around condenser inlet (a) and vacuum take-off (b) will help keep it in place) and attach a 50-ml receiving flask to this adapter.

Attach the lower end of the water jacket of the condenser to the water line with rubber tubing and the upper end to the drain. Add 1–2 *small* (3–4 mm) pieces of clay plate or "Boileezers" to the flask and insert the thermometer.

Have your apparatus checked by your instructor.

A. DISTILLATION OF A PURE LIQUID

Pour 25 ml of 2-propanol into the distilling flask using a long stem funnel so that the liquid will not run down the side arm. Add a "Boileezer" to the flask, replace the thermometer, start the water circulating gently through the condenser (a rapid flow is unnecessary), and check all glass-to-glass connections to make

[1]Fischer Scientific Company.

FIGURE 3.1 A simple distillation assembly using standard-taper glassware.

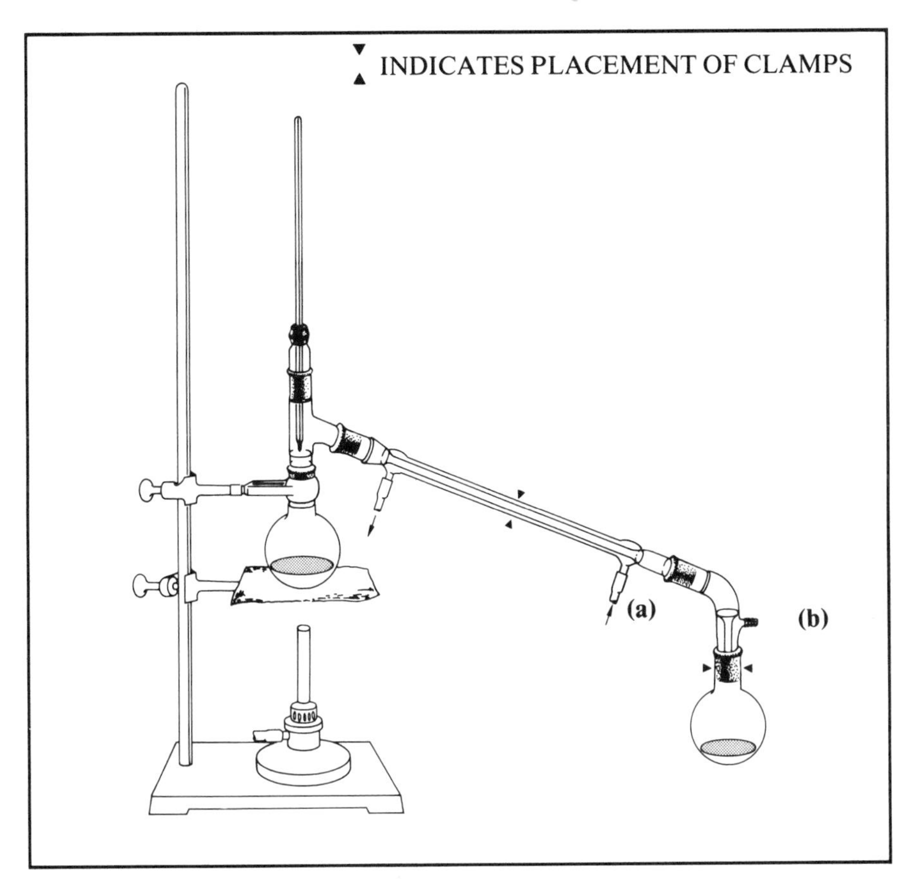

certain they are tight. Adjust your burner to a low flame and gently heat the flask with the burner until the liquid begins to boil. **CAUTION!** *2-Propanol is flammable. Make certain your burner is a safe distance from your neighbor's receiver, or lead vapors away from work area as illustrated in Figure 19.1.* When the liquid begins to drip into the receiver, adjust the burner so that the distillate drops at a rate of about one drop every two seconds. Record (on the report form) the temperature of the vapor when the first drop of distillate falls into the receiver. Continue to collect the distillate until all but 1–2 ml has distilled. Record the temperature of the vapor again and extinguish the burner. The two temperatures recorded define the observed boiling point. Return the distillate to the bottle provided on the side shelf. Discard the residue in the distilling flask.

B. DISTILLATION OF AN AZEOTROPE

Repeat procedure (A) using 30 ml of 25% ethyl alcohol. **(CAUTION!** *Ethyl alcohol is flammable.*) Remove the small flask used previously as a receiver and collect the distillate in a 25-ml graduate. Record the temperature of the vapor at the first drop and at the end of each 5 ml as the distillation proceeds until 25 ml of distillate has been collected. Record the temperatures on the report form.

TEXT REFERENCE

Report: 3

Chapter Pages

Section Desk NAME

DISTILLATION

SIMPLE DISTILLATION OF A PURE LIQUID

Volume collected	Temperature, °C
First drop	
20 ml	
Observed boiling point	
Boiling point (literature)	

DISTILLATION OF AN AZEOTROPE

Volume collected	Temperature, °C
First drop	
5 ml	
10 ml	
15 ml	
20 ml	
25 ml	

Note. Questions and exercises for this experiment are combined with those for Experiment 4.

Experiment 4

Time: 2 hours

Fractional Distillation

A mixture of miscible liquids — i.e., each totally soluble in the other but boiling at different temperatures — cannot be separated satisfactorily by simple distillation. The first distillate of such a mixture (called the forerun), although containing some of the higher boiling material, will always be richer in the lower boiling component. If the initial distillate is redistilled, the first vapor and condensate again will be richer in the lower boiling component. The process, if repeated a great many times, will result in a fairly clean separation of the mixture. Obviously, this would be a laborious and time-consuming process. A **fractionating column** is an ingenious device which eliminates the necessity for this multiple manipulation. The fractionating column (Figure 4.1) is a vertical column packed with some inert material such as glass beads, glass helices, or clay chips to provide a large surface upon which vapor can condense. As the hot vapor rises through this packing it condenses in the cooler part of the column. The condensate flows downward until it reaches a portion of the column sufficiently hot to reconvert it to vapor. Each time the condensate vaporizes, the vapor is again richer in the lower boiling component than the preceding portion, and the residual liquid in the still pot becomes richer in the higher boiling component. This process, repeated a great number of times in the packed columns, finally produces vapor of the lower boiling component that passes as pure compound through the side arm of the column and into the condenser. The lowest boiling component will continue to pass over at its boiling point until it is almost completely separated from the mixture. The temperature of the liquid mixture remaining in the flask then will rise to the boiling point of the next lowest boiling component, and so on until a separation of the liquid mixture is made. This process is called **fractional distillation.** The temperature-composition diagram in Figure 4.2 shows graphically what the fractionating column accomplishes. A mixture of acetone and water in a ratio of approximately 1:3 (composition C_1 in diagram) will produce a vapor whose composition V_1 condenses to liquid of composition C_2.[1] The liquid of composition C_2 produces a vapor of composition V_2. Vapor V_2 condenses to liquid of composition C_3, etc., until finally at point Y on the diagram pure acetone distills.

[1]The vapor above a two-component mixture generally will be richer in the lower-boiling component than the liquid. Thus, *liquid* of composition C_1 will have above it *vapor* of the composition represented by the vertical line drawn through V_1 and C_2 to the composition axis and will boil at temperature indicated by the horizontal line drawn through C_1 and V_1 to the temperature axis.

FIGURE 4.1 A fractional distillation assembly using standard-taper glassware.

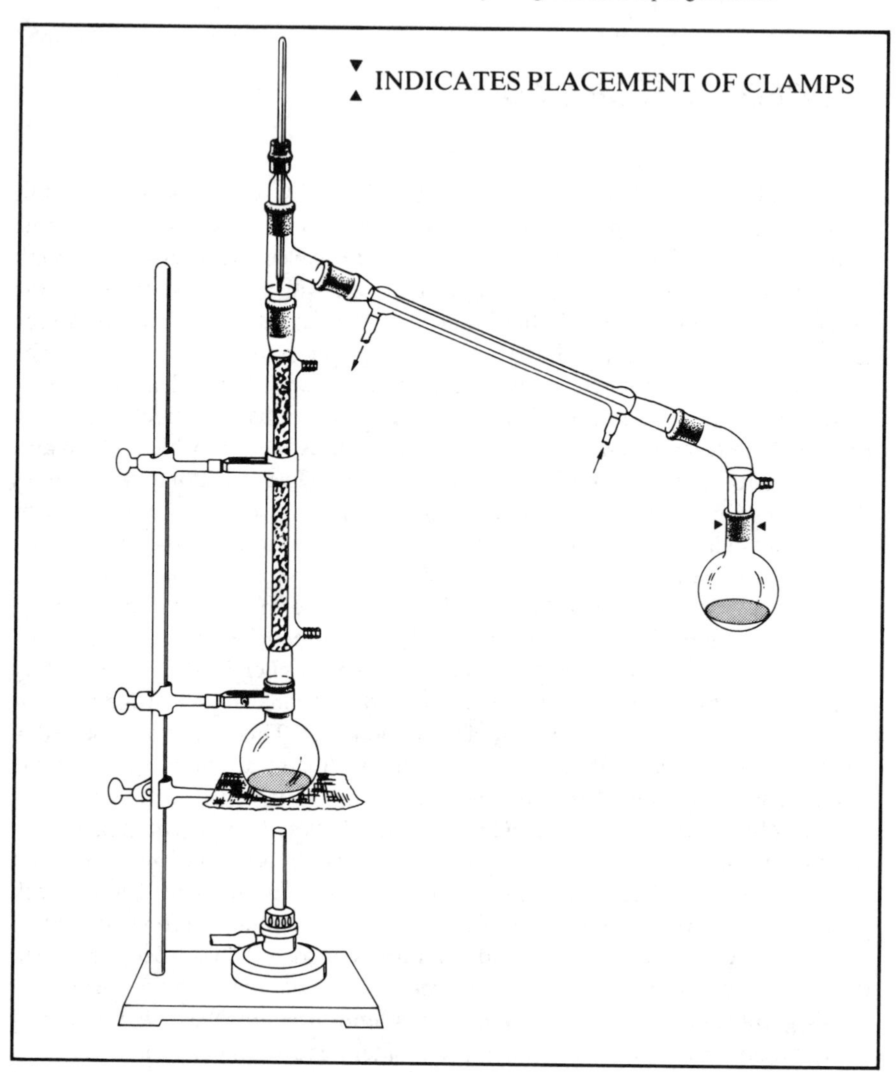

In general the smaller the difference between the boiling points of the two liquids in a mixture, the closer together and more nearly horizontal the liquid and vapor curves will be. Thus, there will be a larger number of steps in the schematic representation of the fractionation process, as shown in Figure 4.2(b). Obviously a more *efficient* fractionating column would be required to separate two liquids having a small difference in boiling point than that required to separate two liquids having a large difference. The efficiency of any given column can be indicated by specifying the number of *theoretical plates* the column possesses, where a theoretical plate may be regarded as that portion of a col-

FIGURE 4.2 Composition-boiling point diagram for (a) mixtures of acetone and water and (b) a hypothetical mixture of liquids X and Y differing only moderately in boiling point.

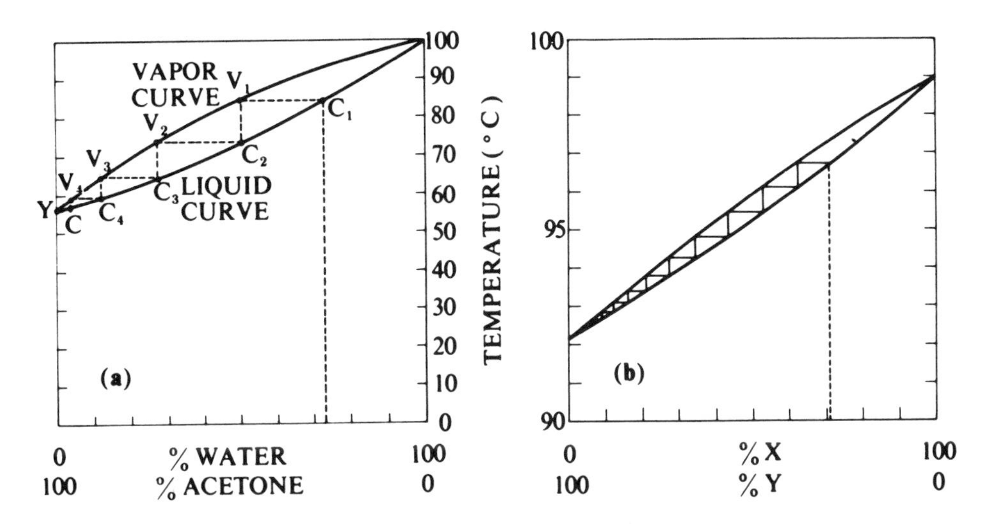

umn's length required to accomplish one step in the fractionation process. For example, a simple distillation set-up (Fig. 3.1) corresponds to roughly one theoretical plate because it can bring about only one step in the fractionation process. Laboratory columns can be constructed having as many as one hundred or more theoretical plates. Although columns having a very large number of theoretical plates are very efficient and can separate liquids boiling only a few degrees apart, the operation of such columns is time-consuming and requires careful attention to operating variables. Simple fractionating columns like that used in this experiment have at best only a few theoretical plates and are not very efficient, but they suffice to illustrate the principles of fractional distillation. Columns not much more complex than that shown in Figure 4.1 are in routine use in organic chemistry laboratories.

Procedure

A. DISTILLATION OF A LIQUID PAIR (WITHOUT FRACTIONATING COLUMN)

Set up a distillation apparatus as shown in Figure 3.1 using a 50-ml round-bottomed flask as illustrated. Introduce 30 ml of a 50-50 mixture of acetone and water into the distilling flask. Replace the flask normally used as a receiver with a 10- to 25-ml graduated cylinder to facilitate recording of the volume of distillate collected. Add several "Boileezers," arrange the thermometer to bring the mercury bulb to the correct position below the side arm of the adapter, and bring the mixture to a gentle boil. Note and record the temperature at which the first drop of distillate appears. If necessary, adjust the heat during the distillation so that the distillate drips slowly and steadily into the graduated cylinder. Record the temperature every 2 ml as the distillation proceeds until the temperature reaches 99–100°. Record the volume of the residual liquid in the distilling flask. On the graph provided on the report forms make a plot of temperature vs. volume collected.

B. DISTILLATION OF A LIQUID PAIR (WITH FRACTIONATING COLUMN)

The apparatus employed for fractional distillation is a modification of that used for simple distillation. Set up a fractionation assembly like that illustrated in Figure 4.1 using a 50-ml round-bottomed flask, a fractionating column packed with glass beads or some other inert material and a thermometer. Do not permit the thermometer bulb to come into contact with the packing material of the column, and make certain the bulb is in the correct position below the side arm opening of the column. For this experiment replace the flask normally used as a receiver with a 10- to 25-ml graduated cylinder as in Procedure (A).

Introduce a fresh 30-ml mixture of equal volumes of acetone and water into the flask as in Procedure (A). Add a "Boileezer" or two, assemble the apparatus, and bring the mixture to a gentle boil. Note and record the temperature at which the first drop of distillate appears. Record the temperature every 2 ml as the distillation proceeds. As necessary, adjust the heat so that the distillate drips slowly and steadily into the graduated cylinder. Continue distilling until all but 1-2 ml of liquid in the flask has been distilled.

On the graph provided on the report sheet make a plot of temperature vs. volume collected.

TEXT REFERENCE

Report: 4

Chapter Pages

Section Desk NAME

DISTILLATION OF A BINARY MIXTURE

Fraction	First drop	1	2	3	4	5	6	7
Volume (ml)	0	2	4	6	8	10	12	14
Temperature (°C)								

Fraction	8	9	10	11	12	13	14	Residue
Volume (ml)	16	18	20	22	24	26	28	
Temperature (°C)								

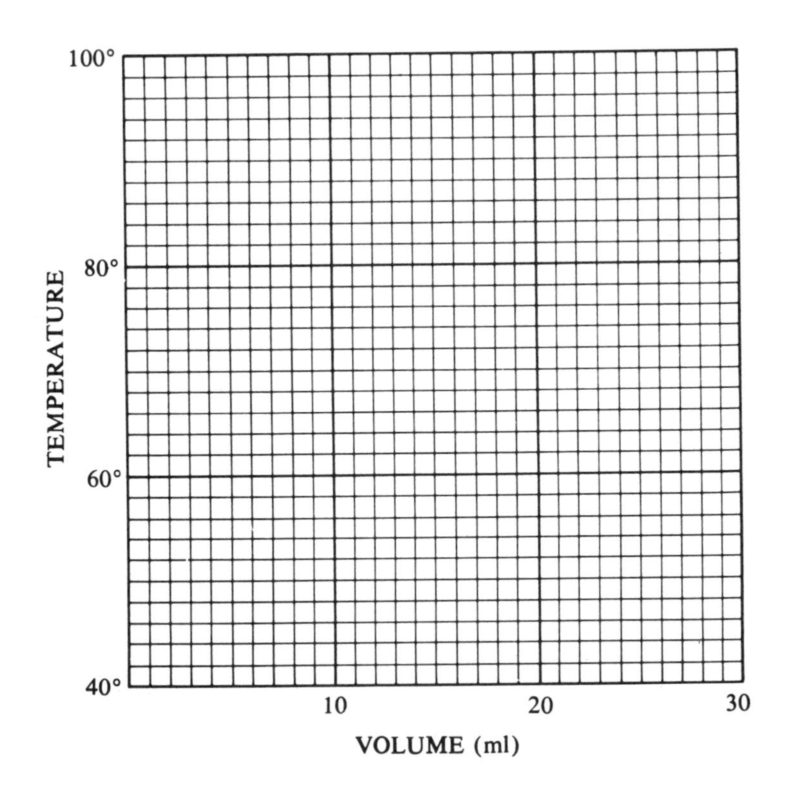

FRACTIONAL DISTILLATION OF A BINARY MIXTURE

Fraction	First drop	1	2	3	4	5	6	7
Volume (ml)	0	2	4	6	8	10	12	14
Temperature (°C)								

Fraction	8	9	10	11	12	13	14	Residue
Volume (ml)	16	18	20	22	24	26	28	
Temperature (°C)								

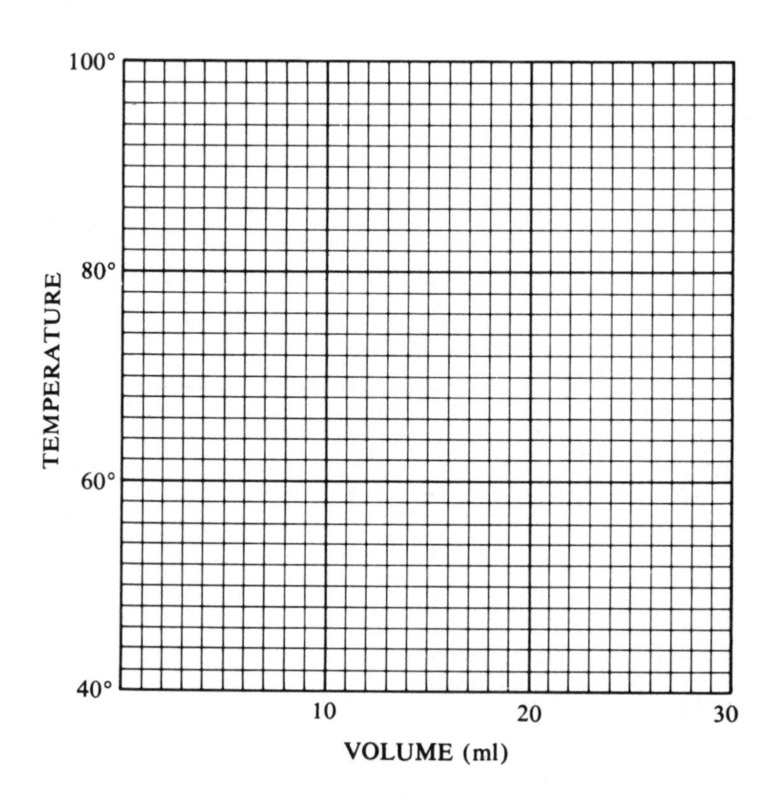

Questions and Exercises

1. If a distillation is started without a "Boileezer" in the distillation flask, the boiling chip should *never* be added to the hot liquid. Explain.

2. In a beginning organic chemistry laboratory in Denver, Colorado (elevation 5,280 ft), two students distil a pure organic liquid, each student taking his sample from the same bottle. One student reports a boiling point 10° below that recorded in the chemistry handbook; the other reports a boiling point 10° above that in the handbook. One student has reported a proper boiling point; the other has not only reported an improper boiling point but also has probably made a common experimental error. Which student has reported the right boiling point; why? Which student has reported the wrong boiling point; what was his experimental error?

3. The efficiency of any column packing material can be expressed in terms of its H.E.T.P. (height of an equivalent theoretical plate) rating, where H.E.T.P. can be defined as that length of column packing corresponding to one theoretical plate. Column packings are known with H.E.T.P.'s of less than 1 cm; simple packings like that employed in Expt. 4 may have H.E.T.P.'s of 10 cm or more. Assume that your packing has an H.E.T.P. of 4 cm and calculate the total number of theoretical plates in your fractional distillation assembly (the complete equipment set-up as in Figure 4.1). To do this measure the length of the packed section to the nearest centimeter. Some students may turn in answers that will be one theoretical plate low. What is the probable source of their error?

4. It is difficult, if not impossible, to specify exactly how efficient a column will be required to effect the separation of a given pair of liquids, for the answer depends not only on what we mean by "separation" ("pure", 95% pure, or what?) but also on a number of experimental variables. However, we can estimate (in a very approximate way) the number of theoretical plates required by use of a number of empirical relationships that have been devised for this purpose. One simple relationship that has given useful results is the following:

$$n = \frac{(T_1 + T_2)}{3(T_2 - T_1)}$$

   where n = the number of theoretical plates, T_2 = the boiling point of the higher boiling substance (in °K), and T_1 = the boiling point of the lower boiling substance (in °K). This equation is intended for use under ideal conditions of "total reflux" (i.e., the column at equilibrium but no distillate being collected so that all of the condensed liquid flows back down the column), and a purity of distillate of 95% or better, but it can be used to estimate the minimum number of theoretical plates required for less ideal conditions. Using this equation, estimate how many theoretical plates would be required to separate acetone and water. [Be sure to convert boiling points in °C to °K by adding 273.1°.]

5. Pure ethanol boils at 78.4° (760 mm), water at 100.0° (760 mm), and the azeotrope (which contains 95.5% ethanol and 4.5% water) at 78.1° (760 mm). Solutions of ethanol and water containing over 95.5% ethanol can be separated by fractional distillation into pure ethanol and the azeotrope. Solutions containing less than 95.5% ethanol can be separated by fractional distillation into water and the azeotrope. The azeotrope can not be separated into its components by fractional distillation. On the basis of this information how would you expect the vapor-liquid composition curves for ethanol-water to differ from those shown in Figure 4.2? On the graph paper provided draw a reasonable set of vapor-liquid curves for this system. You may exaggerate the difference in boiling point between pure ethanol and the azeotrope to simplify the drawing.

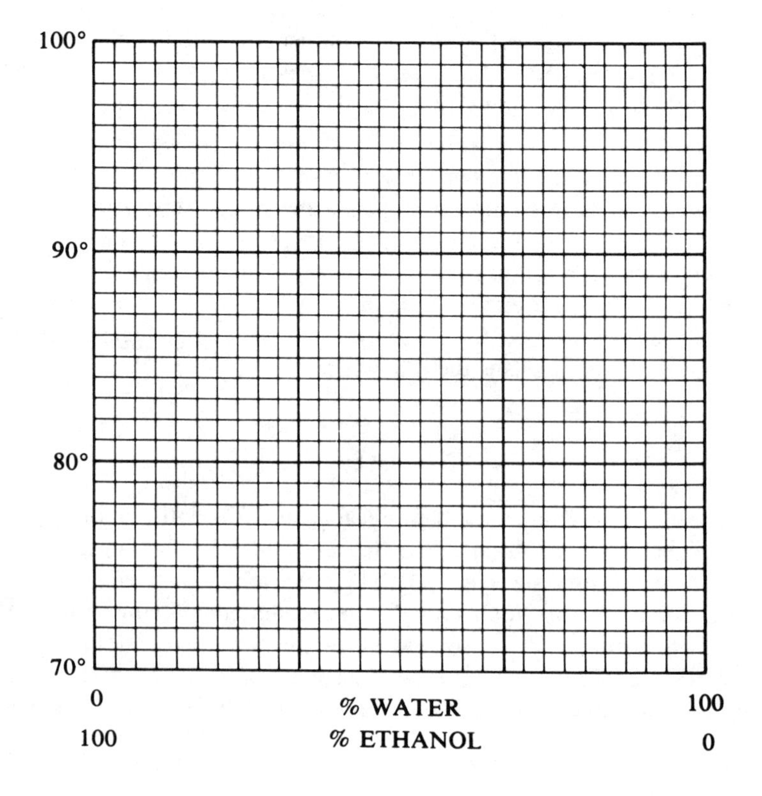

Experiment 5

Time: 3–3½ hours

Extraction

Like recrystallization and distillation, extraction is a separation technique frequently employed in the laboratory to isolate one or more components from a mixture. Unlike recrystallization and distillation, it rarely yields a pure product; thus, the former techniques may be required to purify a product isolated in the crude state by extraction. In the technical sense extraction is based on the principle of the equilibrium distribution of a substance (solute) between two immiscible phases, one of which is usually a solvent. The solvent need not be a pure liquid but may be a mixture of several solvents or a solution of some chemical reagent that will react with one or more components of the mixture being extracted to form a new substance soluble in the solution. The material being extracted may be a liquid, a semi-solid, a solid, or a mixture of these. In fact, extraction is a very general, highly versatile technique which is of great value not only in the laboratory but in everyday life. It is probably one of the oldest chemical techniques used by man.

As was stated above, the substance being extracted may be a solid or semi-solid. Extractions of this type will not be illustrated in this manual, but they are very probably already a part of your own experience. The brewing of tea from tea leaves (or the venerable tea bag which combines extraction and filtration) and of coffee from the ground bean are excellent examples of the extraction of a solid mixture with a hot solvent (water). Other examples include the preparation of vanilla extract from the extraction of the vanilla bean, of gin by the extraction of juniper berries, and of shellac by the extraction of partially purified lac resins obtained from the scale insect, *Coccus lacca*. In each of these examples ethanol is the extracting solvent. Throughout the development of man's technology extraction has been a vitally important process.

In the organic laboratory one of the more important applications of the extraction process has been its use to remove an organic compound from a solution when distillation is not feasible or advantageous. Extraction is accomplished by shaking the solution in a separatory funnel with a solvent that is immiscible with the one in which the compound is dissolved, and one in which the compound dissolves more readily. Two liquid layers thus are formed, and the layer that has most of the desired product in it can be separated from the other. Sometimes not all of the product is extracted in a single operation and the process must be repeated once or twice more to assure a clean separation. It has been found that when two immiscible solvents are shaken together, the solute distributes itself between them in a ratio roughly proportional to its solubility in each. The ratio of the concentration of the solute in each solvent at equilibrium is a constant called the **distribution ratio** (or **distribution coefficient**). For example, at 20° only 0.24 g of azelaic acid will dissolve in 100 ml of water, but 2.70 g of the same acid will dissolve in 100 ml of ether. When shaken with a mixture of equal volumes of water and ether, azelaic acid will distribute between the water and ether layers so that the concentration of azelaic acid in the ether layer will be slightly more than 10 times that in the aqueous layer. The exact value of the distribution ratio, K_d in this case, would be the same as the ratio of the solubilities only if each solvent were *completely* immiscible. This seldom is the case. For practical purposes, however, an *approximate* value of the distribution ratio may be calculated from the following equation.

$$K_d = \frac{C_e}{C_w} = \frac{W_e/100\ \text{ml}}{W_w/100\ \text{ml}} = \frac{2.7}{0.24} = 11.25$$

where C_e and C_w are the concentrations in the ether and water layers respectively, and W_e and W_w are the weights in grams of material dissolved in each respective layer. One can easily calculate the amount of material extracted by a given volume of solvent and how much solute remains in the aqueous layer if the numerical value of the distribution ratio is known. For example, if 0.12 g of azelaic acid in 100 ml of aqueous solution were extracted with 100 ml of ether, the weight (W) of acid extracted by the ether may be calculated as follows:

$$\frac{\frac{W}{100}}{\frac{(0.12 - W)}{100}} = 11.25;\ W = 0.1102\ \text{g}\ (92\%)$$

To determine the amount of material that might have been extracted by using only 50 ml of ether, a similar calculation is made.

$$\frac{\frac{W'}{50}}{\frac{(0.12 - W')}{100}} = 11.25;\ W' = 0.102\ \text{g}\ (83\%)$$

A second 50-ml ether extraction of the residual aqueous solution would remove another proportionate amount of solute according to the distribution ratio.

$$\frac{\frac{W''}{50}}{\frac{(0.12 - 0.102 - W'')}{100}} = 11.25;\ W'' = 0.0153\ \text{g}\ (12.8\%)$$

From these calculations it can be seen that a multiple extraction, using smaller volumes of extracting solvent, more efficiently removes a solute from solution than does one extraction with a much larger volume of solvent.

A second important application of the extraction process in the organic chemistry laboratory involves the use of what is often called a **reaction solvent,** where a reaction solvent can be defined as a *solution* of some reagent (in an appropriate solvent) which reacts selectively with one (or more) components of a mixture to form a new substance(s) soluble in the solvent. To be effective, the reaction solvent must "dissolve" only those compounds which react with the reagent. For example, consider a mixture consisting of an organic acid

$$R{-}C\begin{matrix}\diagup O \\ \diagdown OH\end{matrix}$$

, an organic base or amine ($R'{-}NH_2$), and a neutral organic hydrocarbon ($R''{-}H$), all of which are water-insoluble. The mixture is dissolved in some convenient solvent that is immiscible with water and has a reasonably low boiling point. If the solution of the mixture is first extracted with a dilute (5–10%) aqueous solution of hydrochloric acid, only the amine will react to form the water-soluble organic ammonium chloride.

$$R'{-}NH_2 + HCl \rightarrow R'{-}NH_3^+ + Cl^-$$

Thus, the amine passes from the organic solvent layer into the aqueous layer in the form of the ammonium salt. If the aqueous layer is separated and neutralized with base (e.g., sodium or potassium hydroxide), the insoluble amine will precipitate.

$$R'{-}NH_3^+Cl^- + NaOH \rightarrow R'{-}NH_2 + NaCl$$

Now, if the organic solvent layer (still containing the acid and the hydrocarbon) is extracted with a dilute aqueous solution of sodium hydroxide, only the acid will react to form the water-soluble sodium salt.

$$R{-}C\begin{matrix}\diagup O \\ \diagdown OH\end{matrix} + NaOH \rightarrow R{-}C\begin{matrix}\diagup O \\ \diagdown O^-\end{matrix} + Na^+ + H_2O$$

Thus, the acid now passes from the organic solvent layer into the aqueous layer in the form of the sodium salt. If the aqueous layer is separated and neutralized with acid (e.g., dilute hydrochloric acid), the insoluble acid will precipitate.

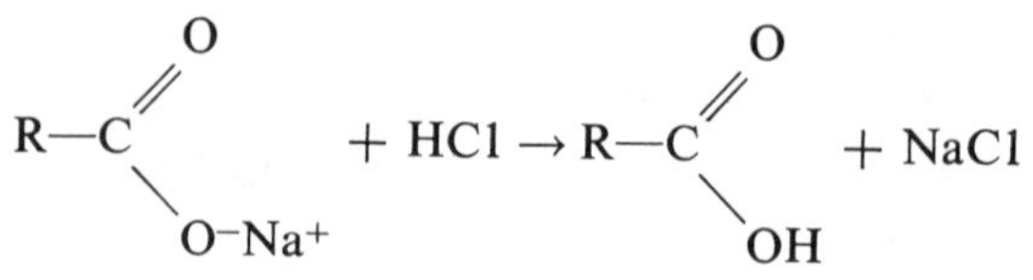

$$R{-}C(=O){-}O^-Na^+ + HCl \rightarrow R{-}C(=O){-}OH + NaCl$$

At this latter stage we have three mixtures: (1) organic amine plus water, (2) organic acid plus water, and (3) a neutral compound in some organic solvent. If either the acid or the amine is solid and highly water-insoluble, it may be recovered by simple suction filtration. However, assume that neither the amine nor the acid can be recovered this simply. Then it will be necessary to extract the water-amine and water-acid mixtures with an appropriate organic solvent to separate these substances from the water. After extractions we have three solutions, each containing one of the three components of the original mixture in an organic solvent. Each component may be recovered by evaporation or, if necessary, fractional distillation of the solvent. Of course, each will probably have to be further purified by recrystallization or distillation.

Sometimes it is helpful in planning a separation scheme based on extraction (alone or with other separation techniques) to prepare a flow chart outlining the various stages in the process. Thus, the example described above could be outlined as follows, assuming that the organic solvent is ether.

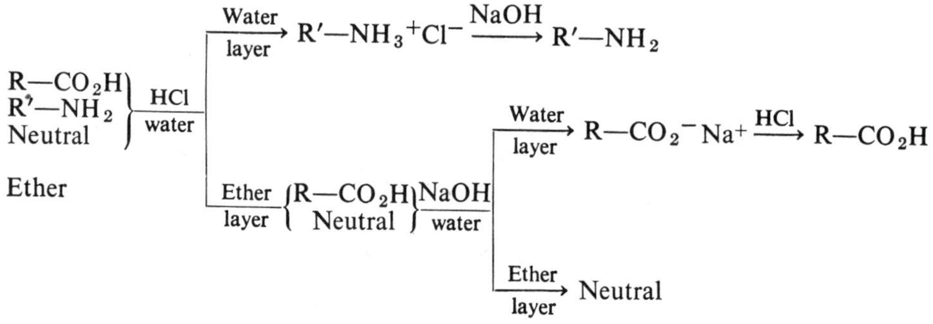

$$\left.\begin{matrix} R{-}CO_2H \\ R'{-}NH_2 \\ \text{Neutral} \end{matrix}\right\} \xrightarrow[\text{water}]{\text{HCl}}$$

Ether

- Water layer → $R'{-}NH_3^+Cl^- \xrightarrow{NaOH} R'{-}NH_2$
- Ether layer $\left\{\begin{matrix} R{-}CO_2H \\ \text{Neutral} \end{matrix}\right\} \xrightarrow[\text{water}]{\text{NaOH}}$
  - Water layer → $R{-}CO_2^-Na^+ \xrightarrow{HCl} R{-}CO_2H$
  - Ether layer → Neutral

From the foregoing discussions some of the desirable properties of an organic extraction solvent become apparent. It must readily dissolve the substance being extracted but must not dissolve to any appreciable extent in the solvent from which the desired substance is being extracted. It should extract neither the impurities nor other substances present in the original mixture. Except in the case of reaction solvents, it should not react with the substance being extracted. Finally, it should be readily separated from the desired solute after extraction. Few solvents will meet all of these criteria and in some cases no completely satisfactory solvent can be found. Therefore, as in the instance of recrystallization, the chemist will select that solvent that most nearly approaches the ideal.

Some of the solvents commonly used for extracting aqueous solutions or mixtures include diethyl ether, methylene chloride, chloroform, carbon tetrachloride, benzene, *n*-pentane, *n*-hexane, and various mixtures of saturated hydrocarbons from petroleum (petroleum ether, ligroin, etc.). Each of these has a relatively low boiling point so that it may be fairly easily separated from the solute by evaporation or distillation. Methanol and ethanol are not good solvents for extracting aqueous solutions or mixtures because of their solubility in water; however, if an aqueous solution can be saturated with potassium carbonate without affecting the solute, ethanol can be used to extract polar solutes from the solution.

FIGURE 5.1 The proper method for handling and supporting the separatory funnel. Notice that the stem of the funnel touches the side of the beaker.

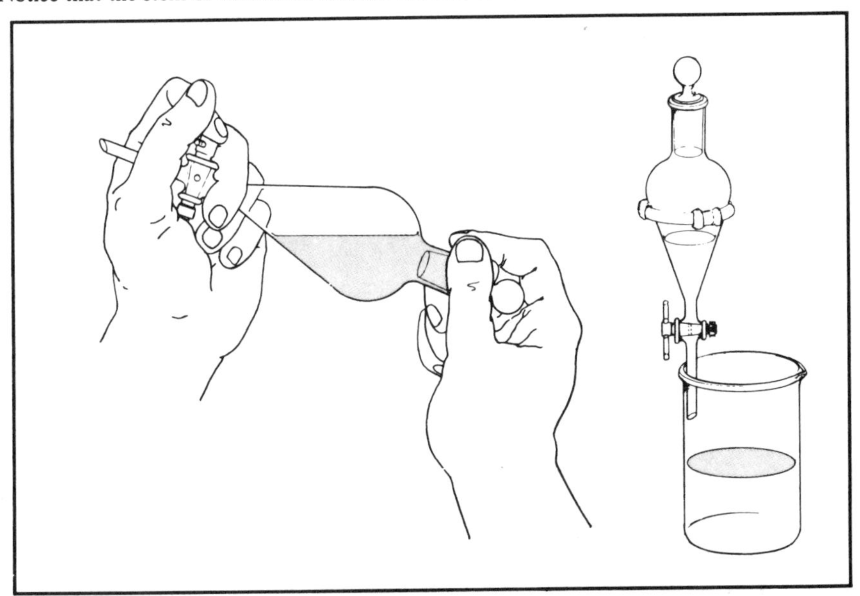

USE OF THE SEPARATORY FUNNEL

The procedure in this experiment involves the use of the separatory funnel. It is important that you learn how to use this rather expensive piece of equipment properly, for, in addition to its use for the extraction of organic substances from aqueous solutions, the separatory funnel is often employed for washing organic liquids. It is made of thin glass and is easily broken unless handled carefully. The following technique for the use of the separatory funnel has been found satisfactory by most laboratory workers.

Support the separatory funnel in a plastic coated iron ring or one padded with short sections of rubber tubing split lengthwise and wired to the ring. Lubricate the stopcock of the funnel (unless it is made of "Teflon," in which case it will require no lubrication). Close the stopcock and add to the funnel the liquids to be separated. Insert the stopper, which should be greased lightly with stopcock grease, and immediately invert the funnel. Point the barrel away from your face (and that of your neighbor) and open the stopcock to release the pressure, which may have accumulated inside the funnel (volatile solvents such as ether develop considerable pressure). The proper method for supporting and handling the separatory funnel is illustrated in Figure 5.1. Close the stopcock and, holding the funnel horizontally, shake the funnel two or three times. Invert the funnel and release the pressure as before. Repeat this process until opening the stopcock causes no further pressure release. Close the stopcock and shake the funnel 15–20 times. Replace the funnel in the iron ring *and remove the stopper.* Allow the liquids to stand until the layers have separated sharply. Draw the lower layer into a flask of proper size. Do not draw the liquid through the stopcock too rapidly. Slow the flow carefully as the boundary between the two layers approaches the stopcock. Stop the flow of liquid completely just as the upper layer enters the hole in the stopcock. Pour the upper layer *through the neck* of the funnel into a second flask. *Never discard either layer until you are absolutely certain which is the proper layer to keep.* Usually one layer will be an aqueous layer or solution, and the other will be an organic liquid. The one of greater density, of course, will be on the bottom.

To check the identity of a layer, should you be in doubt, withdraw a few milliliters of the lower layer into a test tube containing an equal volume of water. If the lower layer in the separatory funnel is water or an aqueous solution it will be homogeneous (only one layer). If the layer being tested is the organic layer, the sample withdrawn will fall to the bottom of your test tube and also form two liquid layers. In either event, return the test mixture to the separatory funnel.

Procedure

A. EXTRACTION OF MANDELIC ACID FROM WATER

In a 250-ml Erlenmeyer flask dissolve 8 g of mandelic acid, $C_8H_8O_3$, in 100 ml of distilled water.

Step 1. Use your graduated cylinder to measure 30 ml of the acid solution and transfer the solution to a 125-ml Erlenmeyer flask. Add 3–4 drops of phenolphthalein and titrate to the end point with a standardized (about 0.3 N) sodium hydroxide solution using one of the buret assemblies provided. Record on your report sheet the number of milliliters of base required to neutralize this volume of acid solution. Calculate the exact weight of mandelic acid dissolved in 30 ml of aqueous solution and also calculate the number of grams of mandelic acid neutralized by each milliliter of the standard base. Discard the neutralized acid solution and rinse your flasks.

Step 2. Measure out a second 30-ml volume of acid solution and transfer it to your separatory funnel. Add 30 ml of ether to the funnel and extract according to the procedure outlined in the introductory section of this experiment. Separate the *lower* aqueous layer into a 125-ml Erlenmeyer flask and add 3–4 drops of indicator. Record the volume of the sodium hydroxide solution in the buret and titrate to the phenolphthalein end point. Again record the number of milliliters of base required and calculate the number of grams of acid removed by the ether and the number of grams remaining in the aqueous layer. Discard the neutralized acid solution and empty the ether layer into the large bottle marked "SOLVENT ETHER FROM EXTRACTIONS."

Step 3. Repeat the procedure from Step 2, but this time extract 30 ml of fresh acid solution with only 15 ml of ether. Separate the aqueous layer into a flask and dispose of the ether layer. Transfer the aqueous layer back into the empty, cleaned, separatory funnel and extract it with a second 15 ml portion of fresh ether. Separate the extracted aqueous layer, add indicator as before and titrate to the end point. Record the volume of standard base required and calculate how much acid remains in the aqueous layer and the total acid removed by the combined ether extracts. Dispose of the ether extracts as directed, clean your separatory funnel, and store your equipment.

B. SEPARATION OF A MIXTURE BY EXTRACTION

In this part of the experiment a mixture containing an acidic, a basic, and a neutral compound is to be separated into the individual components, which will be chosen from the substances whose formulas are shown below.

$C_6H_5—CO_2H$
Benzoic Acid
mp 121–122°

$C_6H_5—CH{=}CH—CO_2H$
trans-Cinnamic Acid
mp 132–133°

$C_6H_5—N(H)—C(=O)—CH_3$
Acetanilide
mp 113°

$C_6H_5—C(=O)—NH_2$
Benzamide
mp 132°

$Br—C_6H_4—NH_2$
p-Bromoaniline
mp 66°

$Cl—C_6H_4—NH_2$
p-Chloroaniline
mp 72°

CAUTION! *Exercise care in working with organic amines, especially with aromatic amines of the types shown above. Avoid unnecessary contact of these materials with any part of the body since they can be absorbed through the skin. Wash off any particles or solutions of these materials that come into contact with the body with generous amounts of soap and water.*

In a 125-ml Erlenmeyer flask dissolve 3.0 g of the three-component mixture provided by the instructor in 25 ml of ether. Pour the solution into your separatory funnel. Rinse the flask with 5 ml of ether and add the ether to the separatory funnel. Extract the organic base (amine) from the mixture using two 25-ml portions of 5% hydrochloric acid (Note 1) and following the procedure outlined in the introductory section of this experiment. Draw off the *lower* aqueous layer in each extraction into the same 125-ml Erlenmeyer flask. Extract the ether layer with 10 ml of water and add the water to the combined HCl extracts. Label the flask **HCl EXTRACT** and set it aside.

Next, extract the organic acid from the mixture, using two 25-ml portions of 5% aqueous sodium hydroxide (prepared by diluting 25 ml of 10% sodium hydroxide solution with 25 ml of water) and one 10-ml portion of water again drawing all of the lower aqueous layers into one 125-ml Erlenmeyer flask. Label the flask **NaOH EXTRACT** and set it aside.

Pour the ether layer in the separatory funnel *through the neck* of the funnel into a 50-ml Erlenmeyer flask and add 1 g of anhydrous magnesium sulfate. Allow the ether solution to stand for 10–15 minutes with occasional swirling. Filter the solution by gravity through a small plug of cotton placed in the bottom of a conical funnel (Note 2) into a 100-ml beaker. Rinse the magnesium sulfate with 1–2 ml of ether, allowing the ether to run into the beaker. Place the beaker in a good fume hood (**NO FLAMES!**) and allow the ether to evaporate. The residue should be your neutral compound. Weigh the compound and determine its melting point and, if directed by the instructor, determine its mixture melting point with some of the pure substance to establish its identity (Note 3).

Cool the HCl Extract in the ice bath and make it basic by adding 10% aqueous sodium hydroxide until the solution is distinctly alkaline to litmus. Chill the mixture thoroughly and collect the released organic base (amine) by suction filtration, using the Hirsch Funnel. Dry the product in air, weigh it, determine its melting point and, if directed, its mixture melting point with some of the pure substance to establish its identity.

Cool the NaOH Extract in the ice bath and make it acidic by adding concentrated hydrochloric acid dropwise until the solution is acid to Congo Red paper (Note 4). Chill the mixture thoroughly and collect the released organic acid by suction filtration, using the Hirsch funnel. Dry the product in air, weigh it, determine its melting point and, if directed, its mixture melting point with some of the pure substance to establish its identity.

Note 1. It is advisable to use moderately dilute solutions of acids and bases for extraction purposes; 5% aqueous solutions appear to be a reasonable compromise. Approximately 5% hydrochloric acid may be prepared by adding 6 ml of concentrated hydrochloric acid to 50 ml of water.

Note 2. The cotton plug should be no larger than the tapered end of a pencil and must fit snugly in the bottom of the funnel but should not be compressed so tightly that the liquid does not flow through freely. It may be advisable to wet the cotton with a small amount of ether before filtering the solution. It is also good practice to decant as much of the solution as possible into the funnel before adding the mixture of solution and magnesium sulfate.

Note 3. All compounds recovered should be dried and weighed and their melting points taken before the end of the laboratory period. If kept uncovered until the next laboratory period before melting points are taken, the organic base may have oxidized.

Note 4. Congo Red changes color in the pH range of approximately 3 to 5, being red in neutral or alkaline solution and blue in acidic solution. It is particularly useful in the organic laboratory because this range is just below that of aqueous solutions of most carboxylic acids. Therefore, Congo Red paper generally remains red in the presence of carboxylic acids but changes to blue in the presence of mineral acids (such as hydrochloric and sulfuric acids). When a solution is "acidified to Congo Red" or "made acidic to Congo Red," just enough mineral acid is added to cause the indicator paper to turn from red to blue. Be sure that the paper is wet by the aqueous solution and not just by the organic layer.

TEXT REFERENCE

Report: 5

Chapter   Pages

Section   Desk   NAME ______________________________

EXTRACTION OF MANDELIC ACID FROM WATER

STEP 1.

Volume of base required to neutralize 30 ml of acid. __________ ml

(a) Number of equivalents of acid in 30 ml of solution.

$$\frac{\text{ml of base}}{1{,}000} \times N\ (\text{base}) = \_\_\_\_\_\_\_\_\ \text{Eq.}$$

(b) Grams mandelic acid in 30 ml of solution.

(1a) × 152 g (M.W. mandelic acid) = __________ g

(c) Grams acid neutralized by 1 ml of base. __________ g

STEP 2.

(a) Number ml of base used. __________ ml

(b) Number of grams of acid remaining in aqueous layer after one 30-ml ether extraction

(2a) × (1c) = __________ g

(c) Number of grams acid removed by one 30-ml portion of ether.

(1b) − (2b) = __________ g

(d) Percent of mandelic acid extracted.

(2c) × 100 ÷ (1b) = __________ %

STEP 3.

(a) Number of ml of base used. __________ ml

(b) Number of grams of acid remaining in aqueous layer after two 15-ml ether extractions.

(3a) × (1c) = __________ g

(c) Total acid removed by both ether extracts.

(1b) − (3b) = __________ g

(d) Percent of mandelic acid extracted.

(3c) × 100 ÷ (1b) = __________ %

SEPARATION OF A MIXTURE BY EXTRACTION

UNKNOWN SAMPLE NO.____

1. Neutral Compound

Weight of neutral compound: __________ g

mp of neutral compound: __________ °C

Mixture mp of compound with ____________________: __________°C

Mixture mp of compound with ____________________: __________°C

Identity of neutral compound: ____________________

2. Acidic Compound

Weight of acidic compound: __________ g

mp of acidic compound: __________ °C

Mixture mp of compound with ____________________: __________°C

Mixture mp of compound with ____________________: __________°C

Identity of acidic compound: ____________________

3. Basic Compound

Weight of basic compound: __________ g

mp of basic compound: __________ °C

Mixture mp of compound with ____________________: __________°C

Mixture mp of compound with ____________________: __________°C

Identity of basic compound: ____________________

Questions and Exercises

1. What are the advantages and disadvantages of using ether as a solvent for the extraction of organic compounds?
2. What volume of an organic solvent must be used to effect a 90% separation in one extraction when only 2.7 g of a certain organic compound dissolves in 100 ml of water? (K = 15.)
3. What percentage of the organic compound could be recovered if two extractions were made, each time using half the volume calculated in Question 2?
4. An organic compound can be extracted from a water layer by an organic solvent more efficiently if the water layer is saturated with an inorganic salt such as sodium chloride. This effect, called "salting out," increases the partition coefficient in favor of the organic compound. Explain.
5. Carboxylic acids (R—COOH) are relatively strong acids and will dissolve in both 5% aqueous sodium hydroxide and 5% aqueous sodium bicarbonate. Phenols (Ar—OH) are weak acids, and most simple phenols will dissolve in 5% aqueous sodium hydroxide (forming the sodium salts, $Ar—O^-Na^+$) but will not dissolve in 5% aqueous sodium bicarbonate (i.e., they do not react with sodium bicarbonate). With this information and that in the introduction to this experiment, work out a flow chart for the separation of a carboxylic acid (R—COOH), an amine ($R—NH_2$), a phenol (Ar—OH), and a neutral hydrocarbon (R—H). Assume that all four compounds are insoluble in water and soluble in ether.
6. Prepare a list of compounds or substances other than those named in the introductory section (not necessarily pure compounds) that are probably isolated or prepared by extraction of some naturally occurring material (plant, animal, insect, petroleum, etc.).
7. In addition to the properties of a low boiling point and immiscibility with water, why is density important to consider in the selection of an extraction solvent? Would an organic liquid with a density of 1.008 g/ml make a good extraction solvent? Why?
8. In Part B of our experiment, why was a solution of sodium bicarbonate not used as a reaction solvent for the extraction of the acid component of the mixture?

Experiment 6

Time: $2\frac{1}{2}$–3 + $1\frac{1}{2}$ + 2–4 + 2–3 hours

Chromatography

Chromatography (Gr., *chroma*, color; *graphein*, to write) is a technique frequently employed by chemists to separate the components of a mixture. In its original application the method was used for the separation of colored substances, but color need not be a required property for compounds to be separated by this method. Colorless compounds may be rendered visible by other means. A number of separation techniques embodying the principles of chromatography have been developed. Four of these along with one or more experiments for each will be described.

PART I COLUMN CHROMATOGRAPHY

A mixture to be separated by **column chromatography** is introduced as a solution into the top of a vertical column packed with some finely divided, inert material already wet with an adsorbed solvent. Enough fresh solvent is added continuously to the top of the column to carry the mixture down and through the supporting material. Each component of the mixture, in descending through the column packing, partitions itself many times between the adsorbed solvent (called the **stationary phase**) and the moving solvent (called the **mobile phase**). The distribution coefficient with respect to these two phases is different for each compound comprising the mixture, and, therefore, each travels through the supporting material at a different rate. Components of the mixture thus become separated within a short time. If the mixture is one of colored substances, compounds appear in the column packing as distinct, colored bands. Continued addition of solvent to the column finally elutes each compound as a colored solution out the lower end and into a receiver. The recovery of each compound as a pure substance then is possible by simply evaporating the solvent. This type of column chromatography is referred to as **partition chromatography.**

When the column packing is wet with the same solvent as that used to elute the mixture, the rate of descent through the column is dependent upon how strongly each component is adsorbed to the column packing. This type of column chromatography is referred to as **adsorption chromatography.** In either case, differences in solubility behavior or differences in adsorption properties are functions of the molecular structures of the different compounds to be separated.

Procedure

SEPARATION OF DYE MIXTURE BY COLUMN CHROMATOGRAPHY

Prepare a small scale chromatography column of the type illustrated in either Figure 6.1(a) or 6.1(b). The column illustrated in Figure 6.1(a) is prepared by stoppering one end of a 25-cm length of 15-mm diameter glass tubing with a size 0 one-hole rubber stopper through which a medicine dropper has

FIGURE 6.1 Simple column chromatography apparatus.

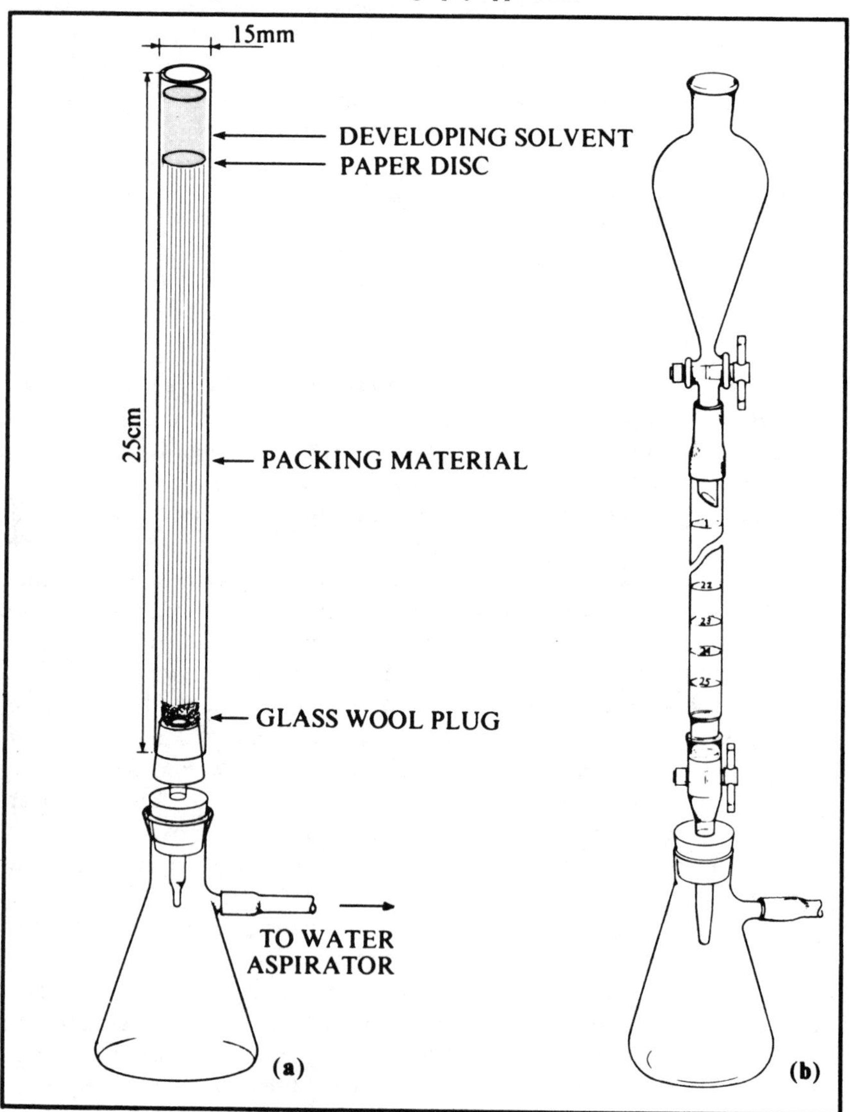

been inserted. The column illustrated in Figure 6.1(b) is prepared from a 25-ml buret equipped with a removable plastic stopcock. Support the column with a buret clamp. Using a length of glass tubing, tamp a plug of cotton firmly into the lower end of the column and cover the plug with a 1-cm layer of sand. Connect the constricted portion of the dropper or the buret delivery tip to a 50-ml filter flask with a tight fitting rubber stopper. Connect the flask to the water aspirator with suction tubing.

Prepare a slurry by mixing 20 g of chromatographic grade alumina, 2 g of "Hyflo Super-cel" or "Celite," and 50 ml of 95% ethyl alcohol. Introduce the slurry into the column through a funnel while tapping the side of the column with a spatula and applying a gentle suction to the filter flask. Fill the column with packing material to within one inch from the top and cover the packing with a disc cut from filter paper. *Never permit the column to be drawn dry of solvent before, during, or after use.* Just as the last of the alcohol is drawn into the packing, add 3–4 drops of a dye mixture prepared by dissolving 5 mg each of Sudan III (oil red) and methylene blue in 5 ml 95% ethyl alcohol. After the sample is drawn into the adsorbent, immediately introduce into the top of the column a developing solvent prepared by mixing 4 volumes of *n*-propyl alcohol, 1 volume of acetone, and 1 volume of water. When fitted with a small separatory funnel as illustrated in Figure 6.1(b), either column may be provided with an ample reservoir of eluting solvent to permit leaving the apparatus unattended for short periods of time. Continue to elute the dye mixture until each dye has been separated and received as a colored solution in the filter flask. The suction on the column may be interrupted after receiving one dye when the eluate immediately following the dye no longer appears colored. When the last compound has made its exit from the column, disconnect the receiver and cap the lower end of the dropper with its rubber bulb and insert a cork into the top of the column or turn stopcocks to closed position. In this way either column may be kept for future use. (Note 1).

PART II PAPER CHROMATOGRAPHY

A mixture to be resolved by paper chromatography is placed as a small spot on one end of a strip or sheet of paper and solvent is allowed to move by capillary action through the spot and up the paper. In this case, the water adsorbed on the cellulose of the paper is the stationary phase. The ratio of the distance traveled by a compound to that traveled by the solvent is called the R*f* value.

$$Rf = \frac{\text{distance the compound traveled}}{\text{distance the solvent traveled}}$$

The R*f* value of a compound is a characteristic of the compound and the solvent used and serves to identify each component of the mixture.

FIGURE 6.2 Spotting for paper or thin layer chromatography. The rule permits holding the paper or film securely. The metric scale on the rule allows you to space your spots uniformly.

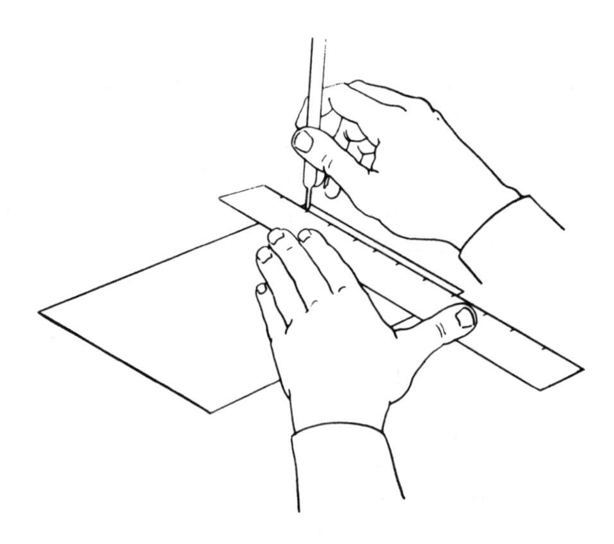

The following experiment illustrates how paper chromatography can be used to separate colored mixtures. The separation and identification of colorless compounds by paper chromatography is illustrated in procedure (H), Experiment 17.

Procedure

THE SEPARATION OF FOOD COLORS BY PAPER CHROMATOGRAPHY

Prepare several micropipettes by heating and drawing out melting point capillaries to a diameter approximately that of a pin or smaller. The bore of a micropipette should be small enough that when it is touched to the surface of a solution or a liquid a 1-cm column of liquid will be retained without a tendency for droplets to form.

Cut a rectangular piece of Whatman #1 chromatography paper measuring 10 cm by 16 cm (see report form p. 61). Draw a light *pencil* line about 1.5 cm from, and parallel to, the edge of the shorter side. Make four pencil marks on the bottom line at intervals of 2.5 cm and identify each mark as R (red) B (blue), Y (yellow), and G (green). Using separate micropipettes for each sample make, in the above order, small (1.5 mm) spots of food colors on each mark (Fig. 6.2). Allow the spots to dry. Roll the paper sufficiently to pass it through the neck of a 500-ml wide-mouthed Erlenmeyer flask. With the lower edge just touching the bottom of the flask, fasten the paper to the inner surface of the neck with transparent adhesive tape. The two vertical edges of the paper must be parallel and must not touch the walls of the flask (see Fig. 6.3). Dilute 10 ml of isopropyl alcohol with 5 ml of water in a small beaker and transfer this solution slowly to the bottom of the flask by pouring it through a long stemmed funnel. Avoid splashing. Cover the mouth of the flask tightly with aluminum foil and allow the chromatogram to develop for about 30–45 minutes. Attach the dry developed chromatogram to your report form.

FIGURE 6.3 Paper chromatography assembly. The filter paper is secured in the Erlenmeyer flask with vertical edges parallel. Lower edge of paper rests on bottom of flask.

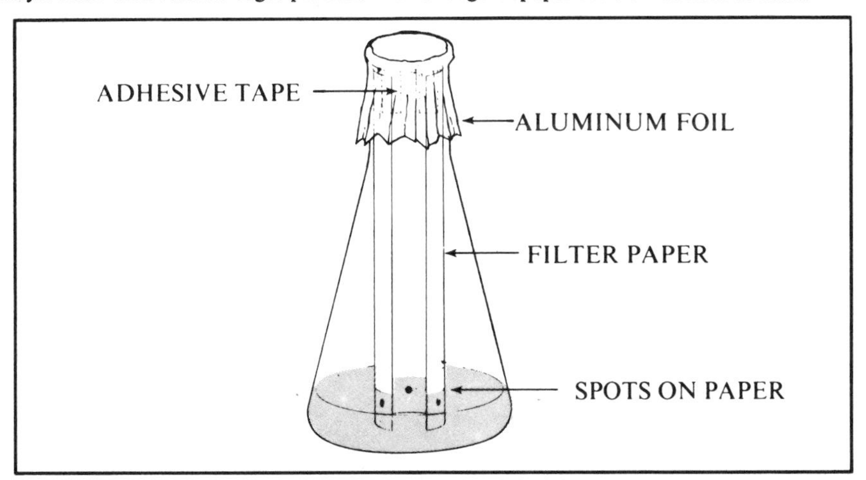

PART III THIN LAYER CHROMATOGRAPHY

Thin layer chromatography, frequently referred to as TLC, is a technique that makes possible a rapid chromatographic separation on thin layers of adsorbent. In practice, a film of approximately 0.3 mm thickness of alumina, silica gel, or other substrate containing a small amount of binding material (usually calcium sulfate) is applied as a slurry to one side of a clean glass plate or plastic film. The slurry then is dried either at room temperature or by heating. Samples of mixtures to be separated are applied as small spots near the bottom of the plate or film, and the chromatogram is developed by following a procedure similar to that followed in paper chromatography. The plate or film is placed in an upright position in a chamber containing a small volume of developing solvent. As the solvent ascends the thin layer by capillary action, components of the mixture become separated and appear as distinct colored spots if the mixture is one of colored substances. If the mixture is composed of colorless substances, individual components may be rendered visible either by the use of ultra-violet light or by chemical means. Unlike paper, the thin mineral layer on the surface of the plate or film can be treated with a variety of reagents including those of a corrosive nature. By marking the solvent front and measuring the distance traveled by each component, R*f* values for individual compounds may be determined as in paper chromatography. In addition to being rapid, the method is very sensitive. Coated glass chromatoplates and coated plastic film for chromatography are available commercially.

Procedure

In the absence of precision apparatus especially designed for coating glass plates with adsorbent or if commercially available coated film for TLC work is not used, substrate may be applied by the following described technique. The plates so prepared may be used instead of the TLC coated film specified in the procedure.

A. COATING MICROSCOPE SLIDES FOR TLC BY SPREADING THE ADSORBENT

Place fifteen to twenty-four 75 × 25 mm microscope slides of uniform thickness in two or three rows, side by side, on a smooth flat surface (Note 2). Fasten the slides with plastic electrician's tape allowing approximately one-eighth inch of tape to lap over the long edge of each outside row (see Fig. 6.4).

FIGURE 6.4 Preparing a number of microscope slides for TLC by spreading adsorbent in a thin layer with a glass rod.

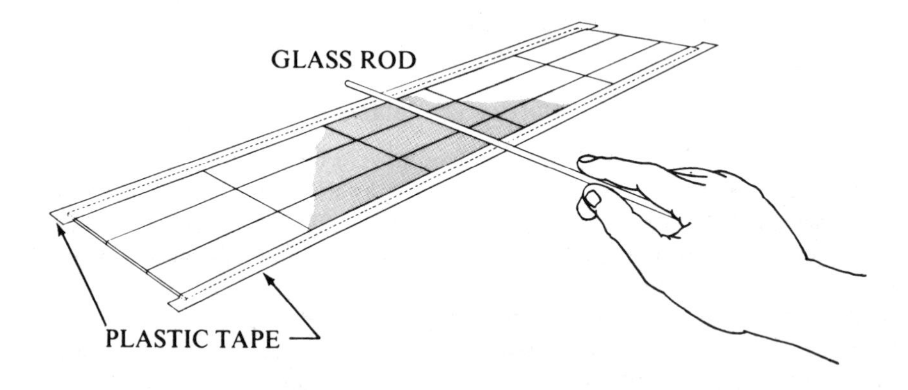

In a 50-ml beaker prepare a slurry by mixing 5 g of silica gel[1] and 13 ml of water to a smooth creamy consistency and transfer it to the slides. Spread the slurry over the slides in a smooth uniform layer by drawing a glass rod back and forth along the tape "rails." Remove the tape, place coated plates on clean pieces of paper toweling, and dry in an oven at 105° or over a hot plate. Store dried plates in a desiccator.

[1]See appendix.

B. THE SEPARATION OF DYE MIXTURES BY TLC

Using a micropipette, apply a small spot (1.0 mm diameter) of a previously prepared dye mixture[1] approximately 1 cm from the end of a 25 × 100 mm strip of coated plastic TLC film. Allow the spot to dry, then lower the film into a 400-ml beaker containing 6 ml of developing solvent prepared by mixing equivalent volumes of *n*-propyl alcohol, acetone, and toluene (Note 3). Cover the top of the beaker tightly with aluminium foil and observe the action on the film. When the solvent has ascended to within 1.0–1.5 cm of the top of the film (approximately 30 minutes), mark the upper limit of the solvent front with a pencil point, remove the film and allow it to dry. Circle each colored oval lightly with a pencil, measure the distance from the midpoint of each colored spot to the starting line, and calculate R*f* values for each dye used in the mixture. Record these values in your report form and compare them with the average values given.

C. THE SEPARATION OF COLORLESS COMPOUNDS BY TLC

Along a line approximately 1 cm from the end of a 50 × 100 mm coated film and 8 mm apart, make four spots of separate samples of phenol, catechol, resorcinol, and pyrogallol using a different micropipette for each. Make, as a fifth spot, a composite of all four compounds. Samples are prepared by dissolving 30 mg of each phenolic compound in 0.5 ml ethanol. Place the film in a 400-ml beaker containing 6 ml of developing solvent prepared by mixing 3 volumes of toluene and 1 volume each of ether, acetone, and ethanol. Cover beaker as before. When the solvent has ascended to within 1.0–1.5 cm of the top of the film (approximately 30 minutes), mark the upper limit of the solvent front, remove the film, and dry. Transfer the dry film into a second clean, dry, 400-ml beaker and add a small crystal (pinhead size) of iodine. Cover the beaker with aluminum foil and warm the bottom over a low flame for a few seconds. Iodine vapor absorbed by the phenolic compounds will reveal the location of each. Carefully outline lightly with pencil each darkened oval area and measure the distance from the midpoint of each spot on the developed chromatogram to the starting line. Determine R*f* values. Record these on your report form and compare with the average values given.

PART IV GAS CHROMATOGRAPHY

Gas chromatography or, more specifically, gas-liquid (phase) chromatography (GLC or GLPC), is a technique employed for the separation and analysis of a mixture of gases, liquids, and even solids if the latter can be vaporized and have an appreciable vapor pressure at temperatures above 100°. To analyze a mixture by GLC a small volume of the mixture is injected into a heated, packed column and the vaporized components are carried through the column by an inert *carrier* gas such as helium or nitrogen.

[1]See appendix.

FIGURE 6.5 Schematic diagram of a gas chromatograph.

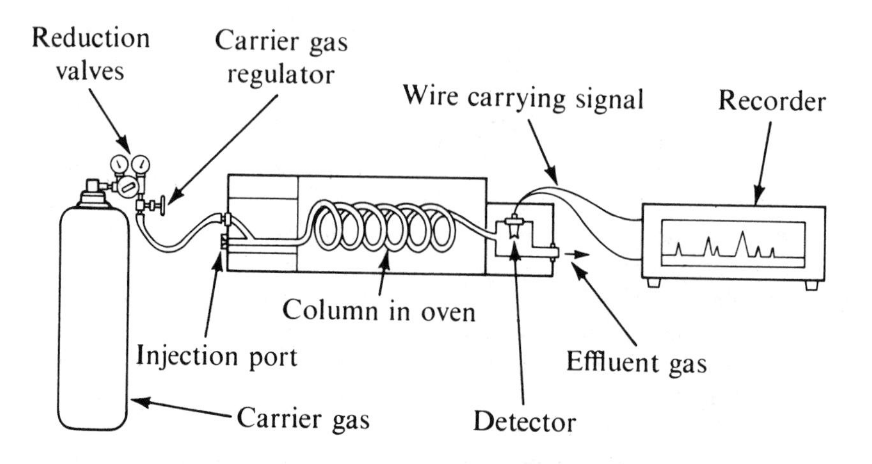

The separation of a mixture by GLC is based on the same principles as are separations by the chromatographic techniques described in Parts I, II, and III, of this experiment, in that the components of the gaseous mixture being carried through the column also become partitioned between a stationary phase and a mobile phase. However, the packed column used in a gas chromatograph, unlike that used in gravity flow liquid-liquid chromatography, need be neither straight nor vertical. The column employed in GLC is often made of a coiled metal tubing of small diameter and is packed with a finely divided, inert, solid material on the surface of which is adsorbed a liquid that is stable and nonvolatile. The former is called the support and the latter the stationary phase. The carrier gas acts as the mobile phase and transports the sample components from the injection point to a detector. The detector monitors the composition of the carrier gas stream as it leaves the column. The principal components of a gas chromatograph are shown schematically in Figure 6.5. A record of the detector response as a function of time is the chromatogram. The chromatogram consists of a base line corresponding to emergence of pure carrier gas and of peaks corresponding to emergence of carrier gas plus sample components. The detector produces a response that is proportional to the concentration of substance in the effluent from the column. A peak area, therefore, is proportional to the amount of a given constituent in the mixture. However, the detector's response may not be exactly the same for two equal concentrations of two different substances. Therefore, peak areas are not necessarily the same for equal amounts of different substances but rather are functions of the individual substances. The composition of a mixture, however, can be calculated from measurements of peak-area ratios. Peak areas may be obtained by use of a mechanical or electronic integrator, by height and width measurements, and even by the simple expedient of cutting out the area and weighing the chart paper on a sensitive balance.

In this experiment we will analyze a mixture of 2-propanol (bp 82.3°) and cyclohexane (bp 81°). As may be seen from their boiling points, it would be difficult to separate a mixture of these two compounds by fractional distillation. However, they are quite easily separated and their percentage composition in a mixture can be determined using GLC.

INJECTION TECHNIQUE

Good results in any gas chromatography experiment are dependent to a great degree upon the injection technique of the experimenter. Care must be exercised to reproduce both the technique and the sample size as accurately as possible. The syringe to be used should be rinsed several times when a new sample is to be analyzed. To rinse the syringe, fill it with sample, expel the contents, and wipe the needle dry with cleansing tissue. To analyze a sample, fill the syringe with a volume in excess of that required, return the plunger to the *exact* volume needed, then retract the plunger to draw in air and wipe the needle dry (Note 4). To inject the sample, center the needle on the septum of the injection block guiding it between the thumb and forefinger to prevent the possibility of bending the needle when it is inserted through the septum. **CAUTION!** *The block is hot!* Insert the needle rapidly, inject the sample, and immediately withdraw the needle from the injection port.

Procedure

Although no general operating procedure can be given for every type of gas chromatographic separation, the procedure that follows is relatively basic and can be applied to any GLC instrument if modified to that instrument's operating instructions. The procedure for this experiment was adapted to a Gow-Mac Model 69–100 gas chromatograph.

Obtain from your instructor 2-ml samples of 2-propanol, a standard mixture of cyclohexane and 2-propanol, and an unknown mixture of these two compounds. Adjust the instrument settings to obtain a flow rate of approximately 60 ml/minute for the carrier gas, a column temperature of 125°, and an injection port temperature of 130°. Balance the recorder. Load approximately 0.5 μl of 2-propanol and 5.0 μl of air into the syringe and inject the sample into the column (Note 5). Mark the chart paper at this point. When the 2-propanol is completely eluted, i.e., when the peak has reached a maximum and returned to the base line, inject a 0.5-μl sample of the standard mixture. It may be necessary here to adjust the recorder attenuation to keep the pen on scale. After both components of the known mixture have been eluted, inject a 0.5-μl sample of your unknown. After your unknown sample has been eluted, tear off your portion of chart paper and write your name on it. Measure the distance (or note the time) between the air peak and the 2-propanol peak on the 2-propanol chromatogram. This distance in time units is called the **retention time** (t_p) for 2-propanol and should be a constant for a given column operated under exactly the same conditions. Measure the retention times of the peaks for the standard sample and also the unknown mixture. With this information identify the peaks that are those of the cyclohexane–2-propanol standard sample (C_s and P_s), as well as those of the cyclohexane–2-propanol unknown mixture (C_u and P_u).

The area (A) under each peak in the chromatogram may be computed by carefully measuring the peak height and its width at the peak's half-height and multiplying these two values (see Fig. 6.6).

FIGURE 6.6 Measurement of the peak area.

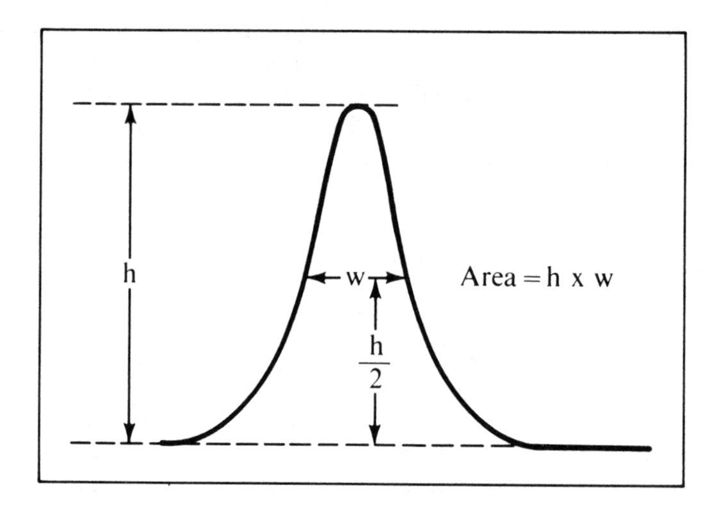

Assume that the area under the peak is proportional to the weight (wt) of substance in the sample. Then $A_P = K_P Wt_P$ and $A_C = K_C Wt_C$.

K_P and K_C are constants; however, $K_P \neq K_C$ because the detector response will probably not be the same for equal concentrations of 2-propanol and cyclohexane.

$$Wt\%C = \frac{Wt_C 100}{Wt_C + Wt_P}$$

$$= \frac{100(A_C/K_C)}{A_C/K_C + A_P/K_P} = \frac{100(A_C/A_P)(K_P/K_C)}{1 + (A_C/A_P)(K_P/K_C)}$$

We do not need to know the value of either K_P or K_C, only a value for the K_P/K_C ratio. This constant ratio (K_r) can be determined from the chromatogram of the standard sample using the formula

$$\frac{K_P}{K_C} = \frac{Wt\%C_s}{Wt\%P_s(A_C/A_P)_s} = K_r$$

The weight % of cyclohexane in the unknown mixture can then be determined from

$$Wt\%C_u = \frac{100\left(\frac{A_C}{A_P}\right)_u K_r}{1 + \left(\frac{A_C}{A_P}\right)_u K_r} = \frac{100K_r}{\left(\frac{A_P}{A_C}\right)_u + K_r}$$

and

$$Wt\%P_u = 100 - Wt\%C_u$$

Note 1. A number of students, if all using aspirators, may reduce the water pressure to the point where very little suction is obtained by any one. The instructor may wish to allow only a few students to perform Part I at one time or he may wish to demonstrate this part of Experiment 6 with students performing Parts II and III.

Note 2. A piece of plate glass 6 × 18 inches or a piece of tempered hardboard of one-fourth inch thickness makes a satisfactory working surface.

Note 3. A 400-ml, high-form Berzelius beaker without pouring spout serves as an excellent developing chamber.

Note 4. Having air in the forward part of the needle prevents vaporization and loss of sample during injection and causes an air peak to appear in the chromatogram that can be used to establish the time base for the chromatogram.

Note 5. The Gow-Mac Model 69-100 is a dual column chromatograph. Column B, containing Chromosorb B, DC-200, must be used for this experiment.

TEXT REFERENCE

Report: 6

Chapter Pages

Section Desk NAME

PART II **PAPER CHROMATOGRAPHY**

THE SEPARATION OF FOOD COLORS

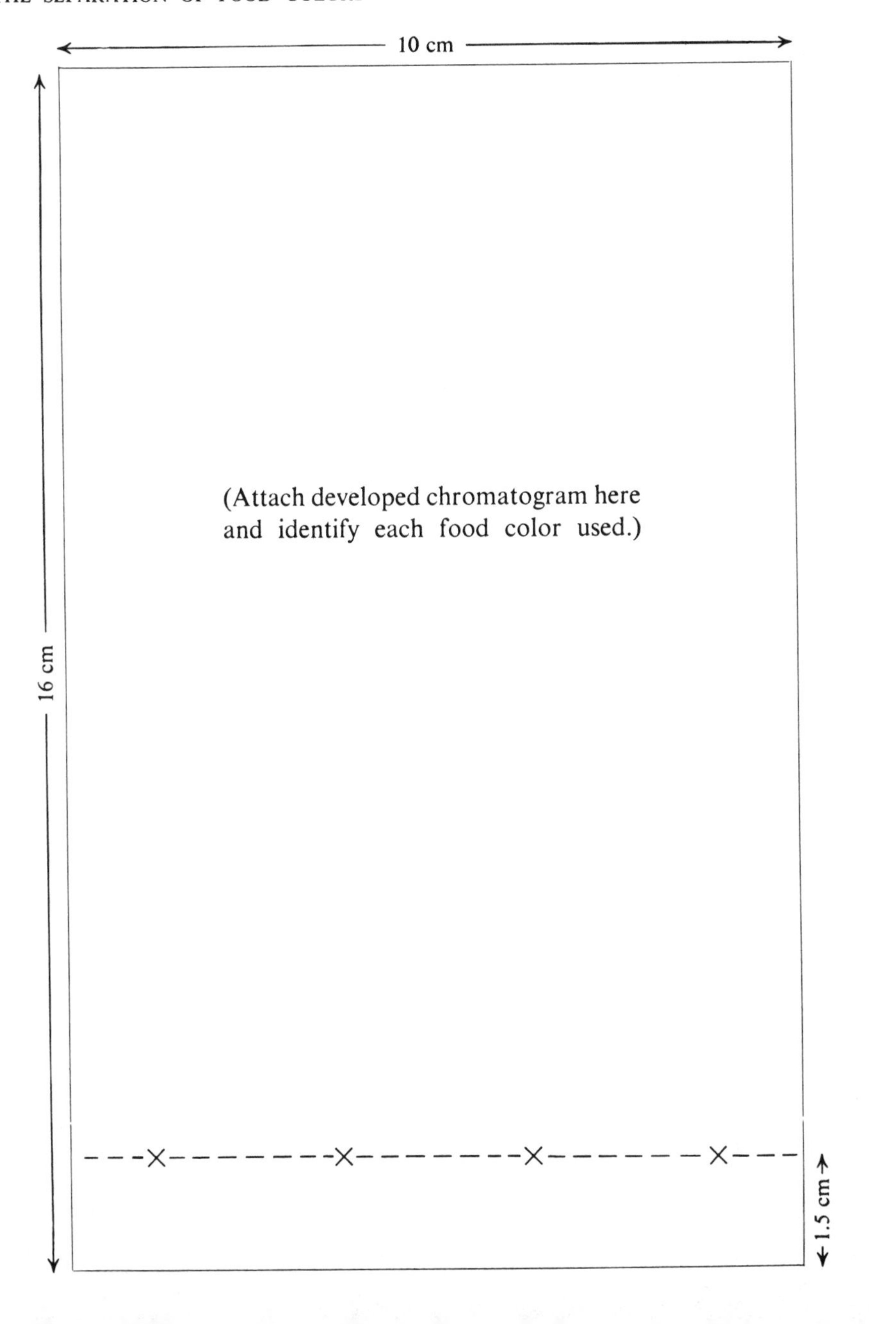

PART III THIN LAYER CHROMATOGRAPHY

THE SEPARATION OF DYE MIXTURES

Mixture	R*f* values (average)	R*f* values (experi-mental)
Butter yellow	0.90	
Rhodamine-B	0.50	
Methylene blue	0.06	

If coated microscope slides were used for TLC dye separation, indicate the location of each dye in your mixture by sketching the developed chromatogram in the outline at the right. If coated plastic film was used, staple your developed chromatogram in the space below.

TEXT REFERENCE

Report: 6

Chapter Pages

Section Desk NAME

THE SEPARATION OF COLORLESS COMPOUNDS

Compound	R*f* values (average)	R*f* values (experimental)
Phenol	0.84	
Catechol	0.72	
Resorcinol	0.77	
Pyrogallol	0.53	

If coated microscope slides were used for TLC separation of phenolic compounds, indicate the location of each compound by sketching the developed chromatogram in the outlined space. If coated plastic film was used, staple your developed chromatogram in the space below.

TEXT REFERENCE

Report: 6

Chapter Pages

Section Desk NAME

PART IV GLC CHROMATOGRAPHY

THE ANALYSIS OF AN UNKNOWN MIXTURE OF CYCLOHEXANE AND 2-PROPANOL

The chromatographic record of your analysis will be a necessary part of the report for Part IV of the chromatography experiments. Include it along with the following information:

Sample Number________

$Wt\%C_s$*________; $Wt\%P_s$*________

t_C________; t_P________

A_{Cs}________; A_{Ps}________

A_{Cu}________; A_{Pu}________

K_r=________

$Wt\%C_u$________; $Wt\%P_u$________

*Percentages to be furnished by the instructor.

Experiment 7

Time: $3\frac{1}{2}$–4 hours

Reactions of Hydrocarbons

PART I ALIPHATIC HYDROCARBONS

The aliphatic hydrocarbons are divided into three classes: (1) the **alkanes,** (2) the **alkenes,** and (3) the **alkynes.** The alkanes have only single bonds in the molecule and are said to be *saturated* hydrocarbons. They are relatively inert. The alkenes have at least one double bond in the molecule, and the alkynes have at least one triple bond. Both the alkenes and the alkynes are said to be *unsaturated* hydrocarbons and both are relatively reactive. The three classes of hydrocarbons react differently with certain reagents, and the differences may be used to distinguish between the three classes in many instances. The following experiments illustrate the reactions of typical saturated and unsaturated hydrocarbons.

A. ALKANES

Obtain 5–6 ml of petroleum ether or "Skellysolve C" from the side shelf. These materials are purified fractions from petroleum and are used industrially and in the laboratory as solvents. "Skellysolve C" has a boiling range of about 88–98° and is probably a mixture of isomeric heptanes. Carry out the following reactions in **dry** test tubes.

(a) BROMINE IN METHYLENE CHLORIDE[1]

Alkanes react slowly or not at all with bromine *in the dark* but in the presence of light may react fairly rapidly according to the equation

$$R—H + Br_2 \rightarrow R—Br + HBr$$

To each of two 2-ml portions of "Skellysolve C" (or petroleum ether) in separate test tubes add 0.5 ml (about 10 drops) of bromine in methylene chloride solution. Shake the tubes well, place one tube in the desk in the dark and expose the other to a bright fluorescent lamp or incandescent bulb, or better still, to direct sunlight for a period of about 15 minutes. Compare the color of the two tubes. Blow your breath *gently* across the mouth of each tube and observe the result. If hydrogen bromide is present, it will combine with the moisture of the breath to form a faint cloud.

[1] **Note to Instructor.** Methylene chloride is not stable to bromine, but will serve as a reagent in the unsaturation test if prepared and stored in a brown glass-stoppered bottle. It is best prepared in small volumes shortly before needed and disposed of after a week or two.

(b) AQUEOUS POTASSIUM PERMANGANATE (BAEYER'S TEST)

Under normal conditions the alkanes are stable (i.e., unreactive) toward chemical oxidizing reagents.

To 2 ml of dilute (0.5%) potassium permanganate solution add a few drops of "Skellysolve C." Shake well and observe the result.

(c) SULFURIC ACID

Alkanes do not react with sulfuric acid under normal conditions.

To 2 ml of concentrated sulfuric acid add 1 ml of "Skellysolve C" and shake vigorously. **CAUTION!** *Do not use your thumb as a stopper.* Note if heat is evolved or if the alkane appears to dissolve.

B. ALKENES

Obtain 5–6 ml of cyclohexene from the side shelf and carry out the following reactions in dry test tubes.

Cyclohexene is a cyclic alkene having the structure [cyclohexene ring]. The reactions of cyclohexene are the same as those of the acyclic or "straight" chain alkenes. It is used in this experiment because it is readily available in pure form and because it has a fairly high boiling point and will not evaporate during your tests.

(a) BROMINE IN METHYLENE CHLORIDE

Bromine *adds* rapidly to unsaturated hydrocarbons to give organic halogen compounds.

$$R{-}CH{=}CH{-}R + Br_2 \rightarrow R{-}\underset{Br}{\overset{H}{C}}{-}\underset{Br}{\overset{H}{C}}{-}R$$

Inasmuch as halogen derivatives of hydrocarbons are colorless liquids or solids, decolorization of a solution of bromine in methylene chloride may be used as a *test for unsaturation*, provided that no other functional group is present that reacts with bromine.

To 1 ml of cyclohexene add 5 drops of bromine in methylene chloride solution, shake the tube and observe the result. Keep adding 5-drop increments of bromine in methylene chloride solution until a total of 50 drops has been added. Does the orange color of the bromine in methylene chloride solution persist, or does your sample continue to decolorize the reagent? Again blow your breath gently across the mouth of the test tube and observe.

(b) AQUEOUS POTASSIUM PERMANGANATE (BAEYER'S TEST)

In the absence of other easily oxidized groups, alkenes react with neutral or alkaline potassium permanganate solution to form *glycols* according to the reaction

$$3\,R\text{—}CH{=}CH\text{—}R + 2\,KMnO_4 + 4\,H_2O \rightarrow$$
(purple)

$$3\,R\text{—}\underset{H}{\overset{OH}{\underset{|}{\overset{|}{C}}}}\text{——}\underset{H}{\overset{OH}{\underset{|}{\overset{|}{C}}}}\text{—}R + 2\,MnO_2 + 2\,KOH$$
(a glycol) (brown)

This reaction is the basis for the Baeyer test for a double bond. Evidence of the reaction with permanganate is the change in color from the purple of the aqueous permanganate solution to the brown of the manganese dioxide precipitate. The glycols produced by the reaction are colorless and water-soluble. This test is specific for the double bond only when no other easily oxidizable groups are present.

To 2 ml of dilute aqueous permanganate solution add a few drops of cyclohexene, shake well and observe the result. If too little alkene is added the permanganate may be reduced only to the manganate ion stage and give a green solution but no precipitate. In this event add a little more of the alkene and shake again.

(c) SULFURIC ACID

Alkenes react with cold concentrated sulfuric acid to give alkyl hydrogen sulfates.

$$R\text{—}CH{=}CH\text{—}R + HO\text{—}\underset{\underset{O}{\downarrow}}{\overset{\overset{O}{\uparrow}}{S}}\text{—}OH \rightarrow R\text{—}\underset{H}{\overset{O\leftarrow\overset{OH}{\overset{|}{S}}\rightarrow O\;/\;|\;O}{\underset{|}{\overset{|}{C}}}}\text{—}CH_2\text{—}R$$

As the hydrogen sulfates are soluble in concentrated sulfuric acid, the observer sees only the alkene appearing to dissolve. Although the alkene appears to dissolve in the sulfuric acid, it should be noted that a reaction rather than true solution is involved.

To 2 ml of concentrated sulfuric acid add 1 ml of cyclohexene in small portions, shake after each addition, and observe the result. Report all your findings on the report sheet.

C. ALKYNES

The number of commercially available alkynes, although increasing, is very limited, and the simplest member of the class, acetylene, C_2H_2, is by far the most common and useful. Although acetylene is available commercially, it is readily prepared from calcium carbide by the following reaction.

$$CaC_2 + 2\,H_2O \rightarrow H{-}C{\equiv}C{-}H + Ca(OH)_2$$

Procedure

Set up an acetylene generator like one of the models illustrated in Figure 7.1 using a 50-ml filter flask or a round-bottomed flask and a 60-ml separatory funnel. By means of a 12-inch length of rubber tubing connect the side arm of the filter flask to a 20-cm length of 6-mm glass tubing which is bent at an angle of approximately 60° one inch from the end. This is the gas delivery tube. Have your instructor check your apparatus.

Acetylene forms explosive mixtures with air over a wide range of concentrations. Do not start this experiment as long as open flames are in close promixity to your workbench. Do not allow acetylene to escape into the room

FIGURE 7.1 Acetylene generators. Generator (a) is shown with delivery tube connected for bubbling acetylene through toluene. Generators (b) and (c) show position of delivery tube for collecting acetylene samples.

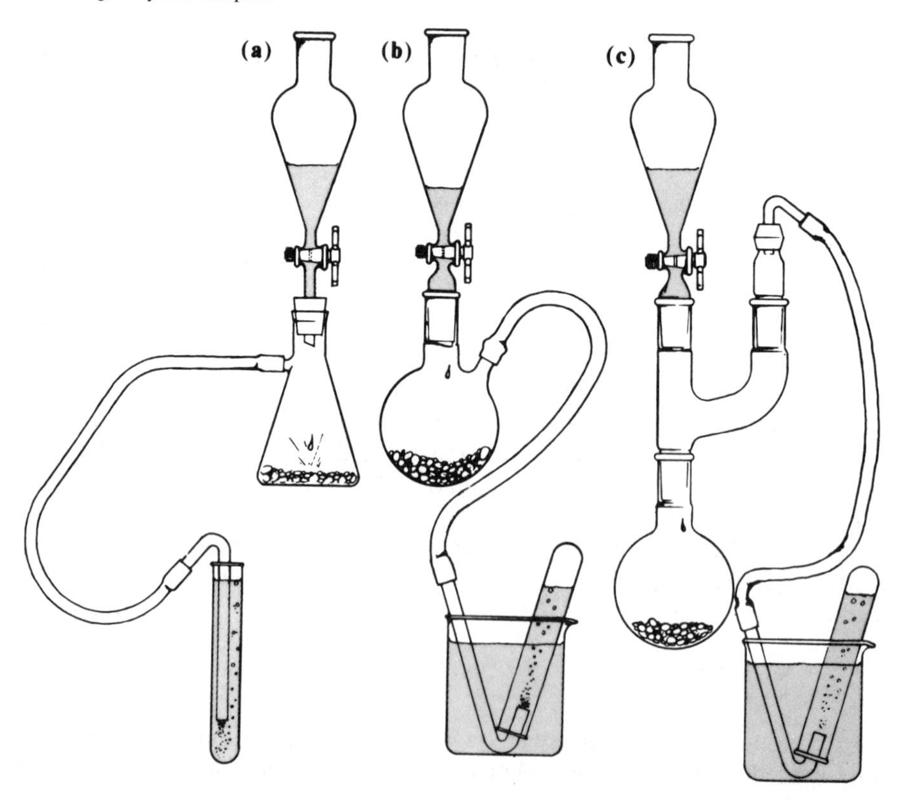

any more than is necessary. When the experiment is finished, destroy any unreacted calcium carbide remaining in your generator as directed at the end of this experiment.

Get all of the reagents which you will require and have them ready. Remove the dropping funnel and place 5 g of calcium carbide in the generating flask, fill the funnel with half its volume of water and replace it. Allow water to drop onto the calcium carbide at a rate no greater than necessary to generate acetylene slowly, and as you need it for the following tests.

(a) FORMATION OF METALLIC SALTS

Acetylene reacts readily with aqueous ammoniacal silver nitrate to give a water-insoluble carbide or acetylide according to the following equation.

$$H{-}C{\equiv}C{-}H + 2\,Ag(NH_3)_2{}^+NO_3{}^- \rightarrow$$

$$Ag{-}C{\equiv}C{-}Ag + 2\,NH_4{}^+NO_3{}^- + 2\,NH_3$$

The silver acetylide is a shock-sensitive explosive when dry.

Only alkynes that have at least one hydrogen on the triply bonded carbon atom show this reaction. Alkenes and alkanes do not react.

Step 1. Prepare an ammoniacal solution of silver nitrate by adding 5–10 drops of concentrated ammonium hydroxide to 5 ml of a 2% solution of silver nitrate. If brown coloration occurs, continue adding NH_4OH drop-wise until a clear solution results. Bubble acetylene through the solution (see part (a) of Figure 7.1) and observe the result. To test the explosive qualities of the silver acetylide collect the precipitate on the Hirsch funnel and dry a **very small** sample on a microspatula by holding it in a flame at some distance from your generator. Destroy the remainder of the silver salt by dissolving it in dilute nitric acid and washing the solution down the drain.

Step 2. Place 5 ml of anhydrous toluene in a clean, dry test tube. Bubble acetylene through the toluene for 2–3 minutes; then add to the toluene solution of acetylene a small piece of clean sodium about the size of a BB shot. Observe carefully. (**CAUTION!** *Any unreacted sodium metal should be destroyed by the addition of 2–3 ml of 95% ethanol to the toluene solution.*) Write an equation for this reaction on your report sheet.

(b) REACTION WITH BROMINE

Bubble acetylene through 2 ml of methylene chloride to which has been added 2–3 drops of bromine in methylene chloride solution. Write an equation for this reaction on your report sheet.

(c) REACTION WITH POTASSIUM PERMANGANATE

Although the reaction of alkynes with potassium permanganate is more complex than that with alkenes, the observed results are the same. Bubble acetylene through 4 ml of dilute (0.3%) potassium permanganate solution in a test tube. Test the solution with blue litmus paper. Record your observations.

(d) FLAMMABILITY

Reverse the glass delivery tube used in procedures (a–c) by connecting the longer end of the bent glass to the rubber tubing [see part (b) of Figure 7.1] and collect a test tube of acetylene by displacement of water. Stopper the sample of gas and take it to the fume hood. Unstopper the tube and touch a lighted match to the open mouth of the test tube. Record your observations.

Fill another test tube with water, but displace only about one-sixth of the water with acetylene. Allow the remainder of the water to slowly drain out. Hold your thumb over the mouth of the test tube for a few seconds. Agitate by repeated inversion, then ignite.

When this experiment is completed, destroy any unreacted calcium carbide remaining in your generator by moving your generator to the hood and adding water until all the carbide is spent. When the evolution of gas ceases, dispose of the contents of the generator as your instructor directs.

PART II AROMATIC HYDROCARBONS

Aromatic hydrocarbons are those which contain one or more C_6 benzenoid structures.

```
          H                          H
          |                          |
          C                          C
        //  \                      /  \\
   H—C        C—H          H—C        C—H
      |        ||     ↔         ||        |
   H—C        C—H          H—C        C—H
        \\  /                      \  //
          C                          C
          |                          |
          H                          H
```

The unsaturation of benzene is due to 6 *p* electrons (one from each carbon atom) in overlapping orbitals which form π bonds between carbon atoms in the ring. The π electrons are not fixed (localized) between any two carbon atoms, but rather are distributed so as to encompass all six carbon atoms within the ring. This overall distribution (delocalization) of electrons confers upon benzene and its derivatives a greater stability than if they were simply cyclic, conjugated trienes.

The reactions of the aromatic hydrocarbons, like those of the saturated hydrocarbons, are principally reactions of substitution.

The following experiments will serve to illustrate some of the properties and some of the more common substitution reactions of toluene—a methyl benzene.

Procedure

A. SOLUBILITY

Test the solubility of toluene in water, ethanol, and "Skellysolve C" (or any other low boiling petroleum ether) using 0.5-ml portions of toluene.

B. FLAMMABILITY

Ignite a few drops of toluene in a small evaporating dish (**HOOD**). Repeat the flammability test using "Skellysolve C". Compare the luminosity of the flame and the amount of soot formed in each case.

C. ACTION OF SULFURIC ACID (SULFONATION)

In a test tube add 1.0 ml of toluene to 8 ml of concentrated sulfuric acid and warm in a water bath maintained at a temperature no higher than 50°. Occasionally remove the tube from the bath, stopper tightly with a rubber stopper, and shake. **CAUTION!** *Keep finger on stopper. Hold test tube away from body.* Continue heating and shaking mixture until two layers no longer separate. Cautiously pour one-half of the reaction mixture onto 10 g of ice. Transfer the aqueous solution to a test tube and examine. Result? On your report form write the equation for the reaction of toluene with concentrated sulfuric acid.

D. ACTION OF NITRIC ACID (NITRATION)

In a 50-ml Erlenmeyer flask prepare a nitrating mixture by cautiously adding 3 ml of concentrated sulfuric acid to 2 ml of 8 *M* nitric acid. Dropwise and with shaking, carefully add 0.5 ml of toulene. After several minutes pour the entire contents of the flask onto 15–20 g of cracked ice. Note the heavy oil which separates. Note its odor. Write the equation for this reaction.

E. ACTION OF BROMINE (BROMINATION)

CAUTION! *One of the reaction products of this experiment is benzyl bromide—a potent lachrymator. Therefore this procedure should be carried out in a good fume hood and a container should be provided in the hood as a receptacle for the disposal of reaction products.*

Place 1 ml of toluene into a 50-ml beaker and 1 ml of toluene in a test tube. Introduce into each of the two samples 2–3 drops of bromine from the bromine buret (**HOOD**). To the test tube sample add a few iron filings or an iron tack. Place the sample in a warm (65°) water bath for 30 minutes. Place the beaker sample directly under a 100-watt incandescent lamp for 2–3 minutes. Blow your breath gently across the mouth of the test tube and the beaker. Observe any color changes. In which sample did a chemical change occur more rapidly? Write equations for all reactions which take place.

F. TEST FOR UNSATURATION

Add 2–3 drops of 0.5% potassium permanganate to 0.5 ml of toluene. Is the color of the permanganate discharged?

TEXT REFERENCE

Report: 7

Chapter Pages

Section Desk NAME

PART I

ALIPHATIC HYDROCARBONS

Reagents used	"Skellysolve C"	Cyclohexene
Br_2 in CH_2Cl_2 (dark)		
Br_2 in CH_2Cl_2 (light)		
0.5% $KMnO_4$		
Conc. H_2SO_4		

ACETYLENE

FORMATION OF METALLIC SALTS

(a) Write an equation for the reaction of acetylene with $AgNO_3$.

(b) What is the color of the precipitate in the above reaction? ________

(c) Did your sample of silver acetylide explode? ________

(d) Write an equation for the reaction of acetylene with sodium.

(e) If the toluene were allowed to evaporate, what would be the nature of the residue? ________

REACTION WITH BROMINE

Write an equation for the reaction of acetylene with bromine.

REACTION WITH POTASSIUM PERMANGANATE

Was the color of permanganate discharged? ______________________

____

What was the effect on litmus? ______________________

FLAMMABILITY OF ACETYLENE

Describe the nature and color of the flame produced by burning acetylene.

____

Was the combustion complete? ______________________

Write an equation for the complete combustion of acetylene.

PART II

AROMATIC HYDROCARBONS

SOLUBILITY OF TOLUENE

Water	Ethanol	"Skellysolve C"

FLAMMABILITY OF TOLUENE AND "SKELLYSOLVE C"

Describe the appearance of the flames. Toluene: ______________________

____

"Skellysolve C": ______________________

Write an equation for the *complete* combustion of toluene.

ACTION OF CONCENTRATED SULFURIC ACID ON TOLUENE

Write an equation for the preparation of toluenesulfonic acid.

ACTION OF NITRIC ACID ON TOLUENE

Write an equation for the nitration of toluene.

ACTION OF BROMINE ON TOLUENE

Write an equation for the reaction of bromine with toluene in the presence of iron filings.

Write an equation for the reaction of bromine with toluene in the presence of bright light.

ACTION OF POTASSIUM PERMANGANATE ON TOLUENE

Is there any evidence of reaction?

Questions and Exercises

1. What is the nature of the gas which evolves when an alkane reacts with bromine?
2. When potassium permanganate solution is decolorized by a sample of a saturated hydrocarbon what conclusions can you draw?
3. Which of the above reactions would you expect to take place with a sample of gasoline? With kerosene?
4. In Step 2 of procedure (A) why was it necessary to use anhydrous toluene?
5. A cylinder of compressed gas has no identifying marking. The gas is flammable and could possibly be either propane or acetylene. Devise a simple test which would identify the gas.
6. Can you think of a reason why acetylene, rather than LP (liquified petroleum gas), is used for welding?
7. Before electricity was readily available, why was acetylene widely used as an illuminating gas?
8. What total volume of acetylene measured at STP could you have prepared from the amount (5 g) of CaC_2 you used in your experiment?
9. What physical property of aromatic hydrocarbons is greatly altered by bonding a sulfonic acid group, —SO_3H, to the aromatic ring?

10. What common physical properties are shared by nearly all nitro compounds? What common chemical property?

11. In each of the procedures (C), (D), and (E) of Part II more than one reaction product is possible. Write the structures and assign acceptable names to all possible products.

12. Draw the structures of T.N.T. and T.N.B. Which of these two high explosives is more easily made? Why?

13. What explanation would you offer if, in procedure (F) of the experiment, the color of permanganate had been discharged? If hot, alkaline $KMnO_4$ had been used on toluene, what would your product have been?

Experiment 8

Time: $3\frac{1}{2}$–4 hours

Steam Distillation: The Isolation of (+)-Limonene from Orange Peel

The distillation of a homogeneous binary mixture (Experiment 4-A) produces a vapor to which each component in the mixture contributes a part. The fraction contributed by each is equal to the mole fraction (N) of the component in the mixture multiplied by the value of the vapor pressure (P°) each component alone would exert at the boiling point of the mixture. This is a statement of Raoult's law, which in equation form appears in every beginning chemistry text as

$$P_{total} = P^\circ_A N_A + P^\circ_B N_B$$

The distillation of a heterogeneous binary mixture, on the other hand, produces a vapor mixture that is not dependent upon the mole fraction of each component present, but only upon the temperature. Thus the total pressure at the boiling point of the heterogeneous mixture is simply the sum of the two separate vapor pressures, or

$$P_{total} = P^\circ_A + P^\circ_B$$

Thus, if a mixture is composed of water and a high-boiling, water-immiscible substance, the mixture will boil near but somewhat below 100°C. Most of the distillate will be water, but a separable portion of it will be the immiscible component, usually in a high state of purity. The process just described, called **steam distillation,** is an excellent technique for the isolation of many water-insoluble compounds that are unstable at temperatures near their boiling points, or compounds difficult to isolate or purify in any other way.

Steam distillations may be carried out in one of two ways. In one method called **direct steam distillation**, the compound to be separated and water are simply placed in a conventional distillation apparatus and the distillate collected. The distillate, which appears as a heterogeneous mixture, may be either two liquid layers or a solid and water. A separation is easily effected by way of the separatory funnel or by a simple filtration. **Indirect steam distillations** are carried out by piping, into a mixture of compound and water, steam generated outside the distillation flask while the distillation flask itself is kept at a temperature near 100°C in order to minimize a build-up of condensate. Apparatus for both methods are illustrated in Figure 8.1. In the present experiment we will isolate by way of direct steam distillation a sample of limonene from one of its many natural sources.

Procedure

Assemble an apparatus as illustrated in Figure 8.1(a). Use a 500-ml flask as the distillation flask and a 250-ml flask as the receiver. Next, with the help of a high-speed food blender, prepare a puree from the parings of four large navel oranges. Use sufficient water to produce approximately 500 ml of slurry (Note 1). Place 300 ml of the orange peel slurry in the distillation flask and bring the mixture to a steady boil with a properly adjusted burner. Avoid violent boiling, for excessive frothing results and may carry small amounts of slurry directly into the condenser. Should this occur, the contents of the receiver must be returned to the distillation flask, the condenser flushed with fresh water, and the distillation

FIGURE 8.1 Assemblies for steam distillation. (a) Direct distillation. (b) Indirect steam distillation.

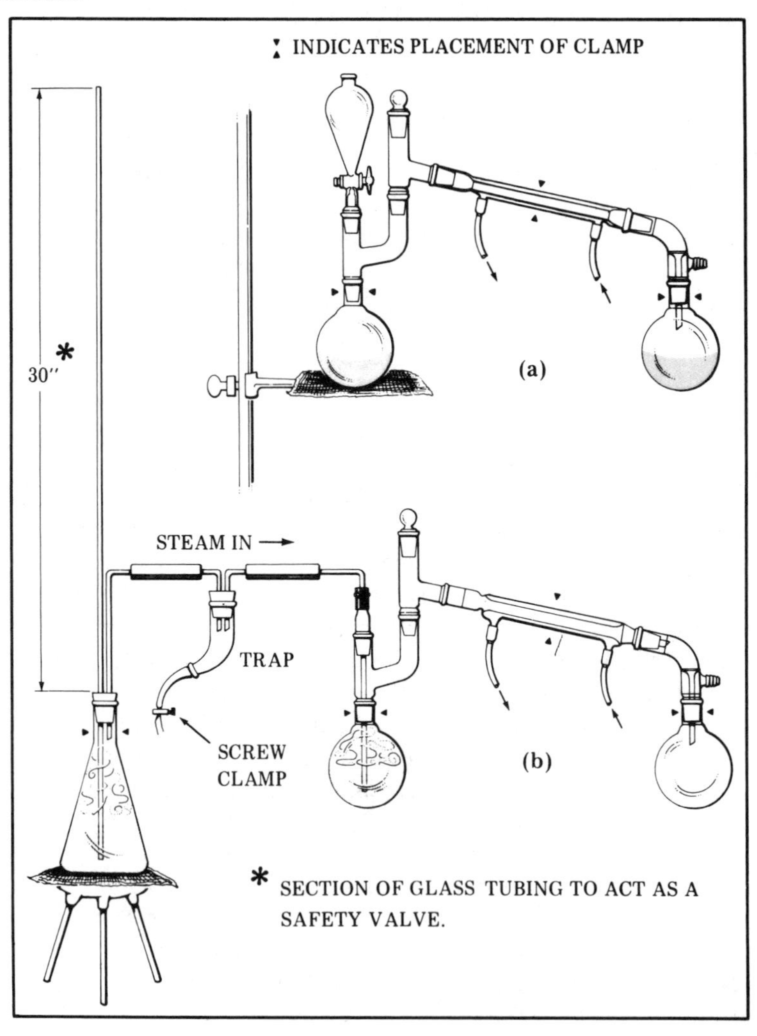

resumed more judiciously. Periodically add 5–10 ml of water to the distillation flask from the separatory funnel to maintain a constant volume. Continue the distillation until about 125 ml of condensate has been collected. By this time an oily layer may be seen on the surface of the collected distillate. If no oily layer appears, it may be necessary to collect a larger volume of condensate or to recharge the distillation flask with fresh slurry. When it is apparent that you have collected some limonene, the distillation may be stopped and any water remaining in the separatory funnel discarded. Transfer the distillate to the separatory funnel and separate the organic layer from the water. Store the limonene collected in a clean test tube over 10–12 granules of anhydrous calcium chloride. The yield of limonene will be 2–4 ml. After the limonene has dried, carry out the following tests (Note 2).

(1) Obtain an ir spectrum of your sample of limonene and compare it to that shown in Figure 8.2. Attach the ir spectrum to your report.

(2) The structure of limonene shows the presence of two double bonds. Test for unsaturation by preparing a solution of 10 drops limonene in 1 ml of methylene chloride and carry out the bromine absorption test as described in Experiment 7, Part I-B.

(3) Although your ir spectrum of limonene may appear to exactly duplicate that shown in Figure 8.2, you cannot be certain that your sample does not contain an impurity. To learn if you have a pure substance, a sample of limonene should be passed through a gas chromatographic column. The general procedure described in Experiment 6-IV may be followed. A column packed with Durapak Carbowax 400 and maintained at a temperature of 150°C and a flow rate of 20 ml/minute will be satisfactory. Record the results of the GLC study on your report form. Also attach the chromatogram.

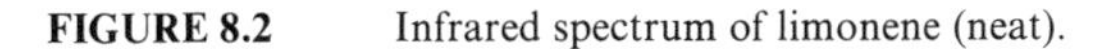

FIGURE 8.2 Infrared spectrum of limonene (neat).

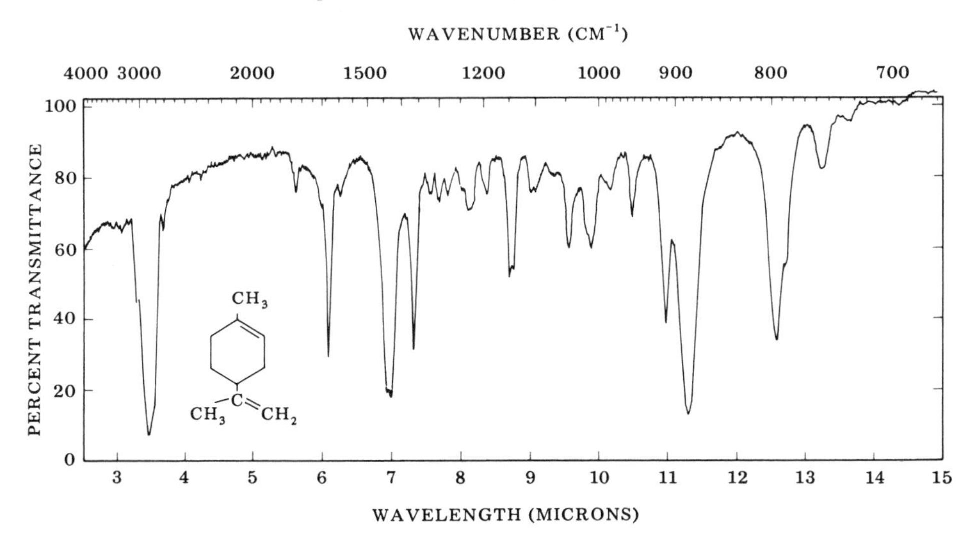

(4) The structure of limonene further indicates that it is a chiral molecule because it possesses a point of asymmetry in a carbon atom bonded to four dissimilar groups. Inasmuch as limonene has its origin in a natural source, we could reasonably expect it to be optically active. Make a solution of 0.75 g limonene in 25 ml of absolute alcohol and fill a 1-dm polarimeter tube with the solution. Set the polarimeter at a zero reading by first viewing the sodium light through a 1-dm tube filled with absolute alcohol. Replace the tube containing the pure solvent with that containing the solution and determine both the direction and extent of the rotation of your sample. Use the average of three measured rotations as your observed rotation measurement. Calculate the specific rotation, $[\alpha]_\lambda^t$, from the following relationship

$$[\alpha]_\lambda^t = \frac{\alpha}{c \times 1}$$

where

α = the observed rotation
t = the temperature at which the measurement was made, normally 20°C
λ = the wavelength of light used. This normally is the D line in the sodium spectrum, a doublet which appears at a wavelength of 5890–5896Å
c = concentration of solution in g/ml
1 = length of light path—i.e., 1 dm

The optical rotation of limonene is reported in the literature as +125°. Record on the report form the specific rotation of your sample of limonene. Return any unused limonene to your instructor in a properly labeled vial.

Note 1. A sharp knife should be used to remove only the outer portion of the orange peel. The heavy white pulp which constitutes the bulk of an orange peel is not to be used.

Note 2. If, at this stage of instruction, the student has not yet reached a level of attainment necessary to understand instrumental methods of analysis, the instructor may wish to do tests (1), (3), and (4) as a demonstration project for the entire class.

TEXT REFERENCE

Report: 8

Chapter Pages

Section Desk NAME

Interpretation of the ir spectrum of limonene:

Type of Bond	Wavenumber (cm^{-1}) and Intensity	Wavelength (microns, μ)	Remarks

Results of test with Br_2–CH_2Cl_2 solution:

Results of GLC study:

Results of polarimetry test: $[\alpha]_D^{20}$ = ________

Yield of (+)-limonene : ________

[1] s=strong; m=medium; w=weak; b=broad

Questions and Exercises

1. To what general class of natural products does limonene belong?
2. There are five structural isomers that possess the limonene skeleton. Draw their structures.
3. Draw structures for the products that could result if limonene were treated with ozone and the resulting ozonide subjected to reductive hydrolysis.
4. The following natural products all may be steam distilled. Which would distill most readily? Why?

Product	Boiling Point
Eugenol	250°
Vanillin	285°
Citronellal	208°

5. Name essences or flavoring agents other than those listed in Question 4 that could probably be isolated by a steam distillation procedure.
6. If your unknown in Experiment 5 (Extraction) had consisted of benzoic acid and *p*-dichlorobenzene, could you have separated your mixture by steam distillation? Why?
7. What properties must an organic compound possess in order to be steam distilled?
8. Would it be correct to say that any organic compound possessing a pronounced aroma is distillable with steam? Explain.

Experiment 9

Time: $2\frac{1}{2}$ hours

A Study of Reaction Rates–The Hydrolysis of tert-Butyl Halides

A large number of organic reactions are those in which one group (a nucleophile, a Lewis base) displaces another. The group displaced is usually a weaker nucleophile and often is referred to as a good leaving group. Such reactions, called substitution reactions, have been studied in great detail with special emphasis upon the factors affecting the rate at which substitutions occur. Important rate-determining factors are: (1) the concentration of the reactants involved in the substitution, (2) the temperature at which the molecules react, (3) the nature of the leaving group and that of its replacement, and (4) the nature of the solvent in which the reaction takes place.

A knowledge of the variables that affect reaction rates is important to the organic chemist for it enables him to carry out a reaction under optimum conditions for the preparation of a product in the greatest possible yield.

The hydrolysis of a tertiary alkyl halide is a chemical reaction that lends itself especially well to a rate study experiment. It is a classic example of a nucleophilic and unimolecular substitution reaction that is commonly designated as the S_N1 reaction. This reaction takes place in two principal steps:

Ionization step (slow step)

$$(CH_3)_3C{-}Cl \rightarrow (CH_3)_3C^{+} + Cl^{-} \qquad (1)$$

Hydrolysis step (fast step)

$$(CH_3)_3C^{+} + H_2O \rightarrow (CH_3)_3C{-}OH + H^{+} \qquad (2)$$

Combining equations (1) and (2) gives the net equation

$$(CH_3)_3C{-}Cl + H_2O \rightarrow (CH_3)_3C{-}OH + HCl \qquad (3)$$

Inasmuch as no reaction sequence can proceed faster than the slowest step, equation (1) is the *rate-determining step* and the speed with which the reaction takes place is dependent upon the concentration of the *tert*-alkyl halide alone. A reaction of this type is said to follow first-order kinetics and the reaction rate may be expressed as

$$\text{rate} = k\,[RX]$$

where k is the reaction rate constant expressed in moles/liter $\times$ sec^{-1}, and [RX] is the concentration of the alkyl halide. The object of the present experiment is to determine a value for k. As may be seen from equation (3), one mole of HCl is produced for each mole of alkyl halide undergoing hydrolysis. By measuring the time required for a predetermined concentration of alkyl halide to undergo hydrolysis, measured by an indicator change when the HCl produced is neutralized by a known volume of standardized base, we will be able to calculate a rate constant for the reaction from the expression

$$kt = 2.303 \log\left(\frac{c}{c-x}\right)$$

where c is the initial concentration of alkyl halide and (c–x) is the concentration after x moles have reacted during time t. The value of k is constant only when the reaction is carried out in the same solvent system and at the same temperature. For this reason it is recommended that burets be filled from stock bottles that have been allowed to equilibrate with room temperature. Also, it is important that reaction times be carefully measured. This is best accomplished by allowing students to work in pairs during this experiment—one mixing reagents, the other carefully noting and recording the time when reaction is begun and completed, along with other data.

Procedure

Part A

Prepare a 0.1 *M* solution of *tert*-butyl chloride in acetone by adding 0.55 ml of *tert*-butyl chloride to a 125-ml Erlenmeyer flask containing 50 ml of *dry*, reagent-grade acetone. Use a 1-ml graduated pipet to accurately measure the alkyl halide. (CAUTION! *Certain of the alkyl halides are cancer suspect. Use a pipet aid to draw up the* tert-*butyl chloride, not your mouth! Avoid contact with skin; avoid breathing of vapor.*) Transfer a portion of the alkyl halide–acetone solution to a 10-ml buret and clamp the buret in place. Fill a 25-ml buret with distilled water. A buret assembly having a white ceramic base is advantageous for this experiment.

Into a clean, dry 50-ml Erlenmeyer flask pipet 0.5 ml of 0.1 *M* sodium hydroxide solution, add 14.5 ml of distilled water from the buret, and add 3 drops of bromphenol blue indicator. Add to the sodium hydroxide solution as quickly as possible 5 ml of *tert*-butyl chloride-acetone solution from the first buret. Swirl the flask while the addition is made to ensure good mixing and start timing the reaction with a stop watch or wrist watch when one-half of the alkyl halide has been added. Record on the report form the total time required for the indicator to change from blue to yellow.

Discard the reaction mixture, rinse your flask with distilled water, and then with a little acetone. Allow the acetone to completely evaporate to dryness before a second determination is made using the same quantities as before. Again note and record the reaction time. The average of the two time values will be taken as t for one-tenth of the *tert*-butyl chloride to undergo hydrolysis.

In the second part of our experiment we will study what effect, if any, a change in the concentration of sodium hydroxide will have upon the rate of hydrolysis of *tert*-butyl chloride. Repeat the previous procedure but this time add 1 ml of 0.1 *M* sodium hydroxide to a 50-ml Erlenmeyer flask containing 14 ml of distilled water and 3 drops of bromphenol blue indicator. Add the same volume of *tert*-butyl chloride–acetone as before. Again note and record the time required for an indicator color change. Repeat and use the average of time values as t for one-fifth of the *tert*-butyl chloride to undergo hydrolysis.

In the third part of our experiment we will study what effect a change in concentration of alkyl halide will have upon the rate of hydrolysis. Prepare a sodium hydroxide solution by adding 1 ml of 0.1 *M* sodium hydroxide to 14 ml of distilled water from the buret. Add 2.5 ml of dry acetone to the basic solution and again 3 drops of indicator. Add 2.5 ml of *tert*-butyl chloride–acetone solution from the first buret. Note and record the time required for an indicator color change. Repeat and use the average time values as t for two-fifths of the alkyl halide to undergo hydrolysis.

Record all concentration and time values on the report form. From these values calculate a value for the reaction rate constant, k.

PART B

A study of structural effects upon the rate of hydrolysis may be carried out by using *tert*-butyl bromide and sodium hydroxide solutions in the same concentrations and volumes as described under Part A, but with the following modification in procedure. Instead of adding the alkyl halide–acetone solution directly from the buret to the sodium hydroxide–indicator solution, the correct volumes of both alkyl halide and base should be contained in 50-ml Erlenmeyer flasks and the temperature of the reactants each brought to 0°C before mixing by suspending the flasks in a slurry of ice water for 5 minutes. The solvolysis of *tert*-butyl bromide at room temperature is so rapid that an accurate value of t is difficult to obtain without a good stop watch.

TEXT REFERENCE

Report: 9

Chapter Pages

Section Desk NAME

THE HYDROLYSIS OF *tert*-BUTYL HALIDE

Part	Vol. NaOH	Vol. RX	$[OH^-]$	[RX]	t(sec)	$\frac{c}{c-x}$	k[1]
I Trial 1							
2							
II Trial 1							
2							
III Trial 1							
2							

Sample calculation (for Part I):

$$kt = 2.303 \log \frac{[RX]}{[RX] - [OH^-]}$$

$$= 2.303 \log \frac{0.025}{0.025 - 0.0025}$$

$$= 2.303 \log 10/9 = 2.303 \log 1.11$$

$$= 0.1043$$

$$k = \frac{0.1043}{t} \frac{\text{moles sec}^{-1}}{\text{liter}}$$

[1] An average value for k is $3.54 \times 10^{-3} \frac{\text{moles sec}^{-1}}{\text{liter}}$ for the solvolysis of *tert*-butyl chloride, and $1.49 \times 10^{-2} \frac{\text{moles sec}^{-1}}{\text{liter}}$ for the solvolysis of *tert*-butyl bromide.

Questions and Exercises

1. What conclusion may be drawn regarding the value of k as a function of concentration of either $[OH^-]$ or [RX]?
2. The solvent in which the hydrolysis of *tert*-butyl chloride took place was approximately 75% water–25% acetone by volume. What effect would increasing the percentage of acetone have upon the rate of hydrolysis? What effect upon the rate of hydrolysis could be expected had we used isopropyl alcohol instead of acetone as the organic component of our solvent? Explain.
3. The intermediate in the S_N1 reaction route is a carbonium ion. What options other than combination with a nucleophile are open to a carbonium ion?
4. If the alkyl halide undergoing hydrolysis via a S_N1 mechanism were an optically pure compound, would the product be optically active? Explain.
5. Experimental evidence shows that complete racemization does not occur when an optically active compound undergoes an S_N1 displacement. Why aren't equal amounts of each isomer produced?
6. List the following halogen compounds in order of reactivity toward S_N1 displacement: (1) *tert*-butyl chloride, (2) allyl chloride, (3) isobutyl chloride, (4) benzyl chloride, (5) *sec*-butyl chloride; (6) neopentyl chloride.

Experiment 10

Time: 2 × 3½ hours

Reactions of the Alcohols and Phenols

The alcohols may be represented by the general formula R—OH. The reactions of the alcohols, if R is a saturated alkyl group, will be mainly those of the relatively reactive hydroxyl group rather than those of the comparatively inert alkyl group.

The phenols, unlike the alcohols, have the hydroxyl group bound to a carbon atom that forms part of the aromatic ring.

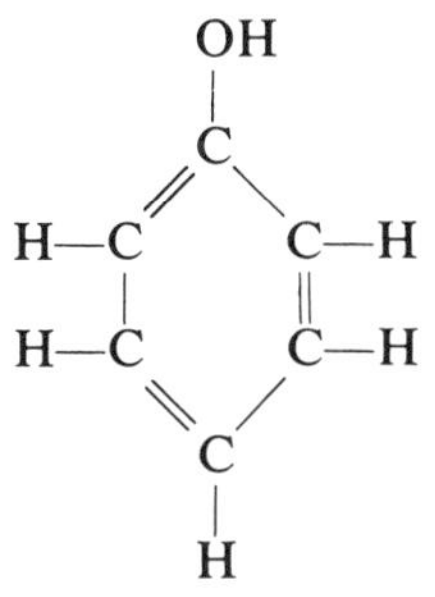

```
         OH
         |
         C
       //  \
  H—C        C—H
     |       ||
  H—C        C—H
      \\    /
         C
         |
         H
```

This feature greatly alters the properties of the phenolic hydroxyl and serves to distinguish it from hydroxyl groups of the alcohols.

The following tests and reactions illustrate some of the general properties of the alcohols and phenols.

PART I PROPERTIES AND REACTIONS OF ALCOHOLS

Procedure

A. SOLUBILITY IN WATER AND IN ACIDIC SOLUTIONS

Into four separate test tubes place small samples (about 10 drops each) of the following: *n*-butyl alcohol, *tert*-butyl alcohol, cyclohexanol, and benzyl alcohol. Add 5 ml of water to each sample, mix, and observe. Next, place all test tubes in a water bath and warm to 65°. Test a drop of each solution on red litmus paper. Remove all samples from the water bath and allow them to cool to room temperature. Observe and record your results.

Repeat part (A) above but instead of water, use equivalent volumes of 85% phosphoric acid. Mix by shaking each solution as before but do not heat. Again observe and record your results.

B. REACTION WITH SODIUM

Water reacts with metallic sodium to form hydrogen and sodium hydroxide. In a similar manner alcohols react with sodium to form hydrogen and sodium alkoxides.

$$2\ R\text{—}OH + 2\ Na \rightarrow 2\ RO^-Na^+ + H_2$$

Phenols react with sodium to yield hydrogen and sodium phenoxide.

$$2\ Na + 2\ C_6H_5OH \rightarrow 2\ C_6H_5O^-Na^+ + H_2$$

Phenol   Sodium phenoxide

Place 2 ml of each of the following alcohols in separate, *dry* test tubes: (1) methyl alcohol, (2) isopropyl alcohol, (3) *n*-propyl alcohol. To each test tube add a small piece of sodium metal (about the size of a BB shot) and observe the result. Allow the reaction to proceed to completion. Record your results on the report form. (**Never put metallic sodium into water!** If necessary, add more alcohol to destroy any excess sodium.)

C. REACTION OF ALCOHOLS WITH LUCAS REAGENT

The *Lucas test* for distinguishing between primary, secondary, and tertiary alcohols (the class of an alcohol is the same as the class of its alkyl group) is based on the relative rates of reaction of the different classes with hydrogen chloride. The Lucas reagent is a solution of zinc chloride in concentrated hydrochloric acid.[1] A tertiary alcohol reacts rapidly with the reagent to give an insoluble alkyl chloride which appears as a cloudy dispersion or as a separate layer in the solution. A secondary alcohol gives a clear solution at first (as the alcohol dissolves) which becomes cloudy *within five minutes.* A primary alcohol dissolves to produce a solution that remains clear for several hours.

$$R\text{—}OH + HCl \xrightarrow{ZnCl_2} R\text{—}Cl + H_2O$$

[1] See appendix.

To each of three test tubes add 1 ml of the Lucas reagent. Add 4–5 drops of the alcohol to be tested, mix well, and record the length of time it takes for the mixture to become cloudy or to separate into two layers. Carry out the tests with: (1) *n*-butyl alcohol, (2) *sec*-butyl alcohol, and (3) *tert*-butyl alcohol. Record your results on the report form.

D. OXIDATION OF ALCOHOLS

When the saturated, aliphatic alcohols are oxidized under relatively mild conditions, the product of the reaction depends upon the class of the alcohol being oxidized. Thus, primary alcohols give first aldehydes and finally carboxylic acids as oxidation products.

$$\underset{\text{Primary alcohol}}{R{-}CH_2{-}OH} \xrightarrow{[O]} \underset{\text{Aldehyde}}{R{-}C(=O){-}H} \xrightarrow{[O]} \underset{\text{Carboxylic acid}}{R{-}C(=O){-}OH}$$

By proper choice of the oxidizing agent it is often possible to stop the reaction at either the aldehyde or carboxylic acid stage. Secondary alcohols give ketones as oxidation products.

$$\underset{\text{Secondary alcohol}}{R{-}CH(OH){-}R'} \xrightarrow{[O]} \underset{\text{Ketone}}{R{-}C(=O){-}R'}$$

Tertiary alcohols are inert toward most mild oxidizing agents.

$$\underset{\text{Tertiary alcohol}}{R{-}C(OH)(R''){-}R'} \xrightarrow{[O]} \text{No reaction}$$

Place 2 ml of each of the following alcohols in separate, clean 50-ml Erlenmeyer flasks: (1) ethanol, (2) isopropyl alcohol, and (3) *tert*-butyl alcohol. To each alcohol sample add 5 ml of an oxidizing solution prepared from 5 g of potassium dichromate in 50 ml of water and 5 ml of concentrated sulfuric acid. Mix the reagents by swirling the flasks and warm on the steam bath.

Observe any changes. Remove the flasks containing ethanol and isopropyl alcohol samples, and place them on a wire gauze over a burner. Bring the mixtures to a boil. Extinguish the burner and, using a tweezer, place over the mouth of each flask a strip of moistened blue litmus paper. After 10 minutes, inspect each flask and record your observations. Write the equations for the reactions (if any) that took place.

E. THE IODOFORM REACTION

Alcohols which have the structure $CH_3—\overset{\overset{OH}{|}}{C}H—R$ (R = H, alkyl, or aryl) and aldehydes or ketones which have the structure $CH_3—\overset{\overset{O}{\|}}{C}—R$ (R = H, alkyl, or aryl) react with an alkaline solution of the halogens (hypohalite solutions) to yield the corresponding haloforms. Indeed, the "haloform reaction" often is performed as a diagnostic test on an unknown compound suspected of having the above structural arrangements. The preparation of iodoform may be illustrated by the following reaction sequence.

$$2\,NaOH + I_2 \rightarrow NaI + NaOI$$

$$R—\overset{\overset{OH}{|}}{C}H—CH_3 + NaOI \rightarrow R—\overset{\overset{O}{\|}}{C}—CH_3 + NaI + H_2O$$

$$R—\overset{\overset{O}{\|}}{C}—CH_3 + 3\,NaOI \rightarrow R—\overset{\overset{O}{\|}}{C}—CI_3 + 3\,NaOH$$

$$R—\overset{\overset{O}{\|}}{C}—CI_3 + NaOH \rightarrow R—\overset{\overset{O}{\|}}{C}—O^-Na^+ + \underset{\text{Iodoform}}{CHI_3}$$

Chloroform and bromoform may be prepared in a similar manner; but, inasmuch as they are liquids, they are less useful for diagnostic purposes. Iodoform, on the other hand, is a yellow solid (mp 119–120°) with a sharp, easily recognizable odor.

To 0.5 ml of acetone (about 10 small drops) in a test tube add 5 ml of water and 1 ml of dilute (6 *N*) sodium hydroxide. To the alkaline solution of acetone add, dropwise, an iodine-potassium iodide solution.[1] Continue to add iodine until the brown color no longer is discharged and only a light yellow color remains. If less than 1 ml of the iodine-potassium iodide solution is

[1]See appendix.

used, heat the test tube in the water bath at 60° and add any additional iodine-potassium iodide needed. Collect the precipitate that forms on the Hirsch funnel, wash it on the filter with 5 ml of (5%) dilute sodium hydroxide, and dry on a piece of clay plate, or on a large piece of filter paper placed on a watch glass. Determine the melting point of the precipitate.

Repeat the iodoform test using (a) ethyl alcohol, (b) methyl alcohol, (c) isopropyl alcohol, and (d) *n*-propyl alcohol (Note 1). The melting point determinations may be omitted. Record your results on the report form.

Note 1. For alcohols having very low water solubility it may be necessary to dissolve the sample (10 drops or 0.1 g) in 5 ml of dioxane. In such instances it is advisable to run a blank on the dioxane, for some samples of dioxane have given positive iodoform tests.

F. DETERMINATION OF AN UNKNOWN

From your instructor obtain an unknown alcohol and determine its identity by a series of tests which will include one or more of the following: the Lucas test, oxidation with dichromate–sulfuric acid reagent, or the haloform reaction. Your unknown will be one of those listed in Table 10.1.

TABLE 10.1

STRUCTURES AND BOILING POINTS (°C) OF SOME ALCOHOLS

Compound	Formula	Boiling point
2-Propanol	$CH_3—CH(OH)—CH_3$	82
2-Methyl-2-propanol	$CH_3—C(CH_3)(OH)—CH_3$	83
1-Propanol	$CH_3CH_2CH_2OH$	98
2-Butanol	$CH_3CH_2—CH(OH)—CH_3$	100

TABLE 10.1 (*Continued*)

2-Methyl-2-butanol	CH_3 $CH_3CH_2—C—CH_3$ OH	102
2-Methyl-1-propanol	CH_3 $CH_3—CH—CH_2OH$	108
3-Pentanol	$CH_3CH_2—CH—CH_2CH_3$ OH	116
1-Butanol	$CH_3CH_2CH_2CH_2OH$	118
2-Pentanol	OH $CH_3—CH—CH_2CH_2CH_3$	119
3-Methyl-1-butanol	CH_3 $CH_3—CH—CH_2CH_2OH$	130
4-Methyl-2-pentanol	CH_3 OH $CH_3—CH—CH_2—CH—CH_3$	132
Cyclopentanol	$CH_2—CH_2$ $CH—OH$ $CH_2—CH_2$	141
2-Methyl-2-hexanol	CH_3 $CH_3(CH_2)_3C—CH_3$ OH	143

A boiling-point determination is a part of this exercise and should be performed as the first step. This is best done by a small scale distillation (Experiment 3) on a sample of 5 ml or more, but, if your sample is but a milliliter or two, may be done as follows.

Make a hole approximately the diameter of a pencil in the center of a 5 × 5-inch wire gauze having a circular asbestos coating. Support the gauze on a ring and position on it a 150-mm test tube with the bottom of the test tube resting directly over the hole (Fig. 10.1). The asbestos shields the outer wall of the test tube from the heat of your burner and allows the tube to act as a condenser. Clamp the test tube securely. Place a 2-ml sample of your unknown in

FIGURE 10.1 Assembly for a small-scale boiling-point determination.

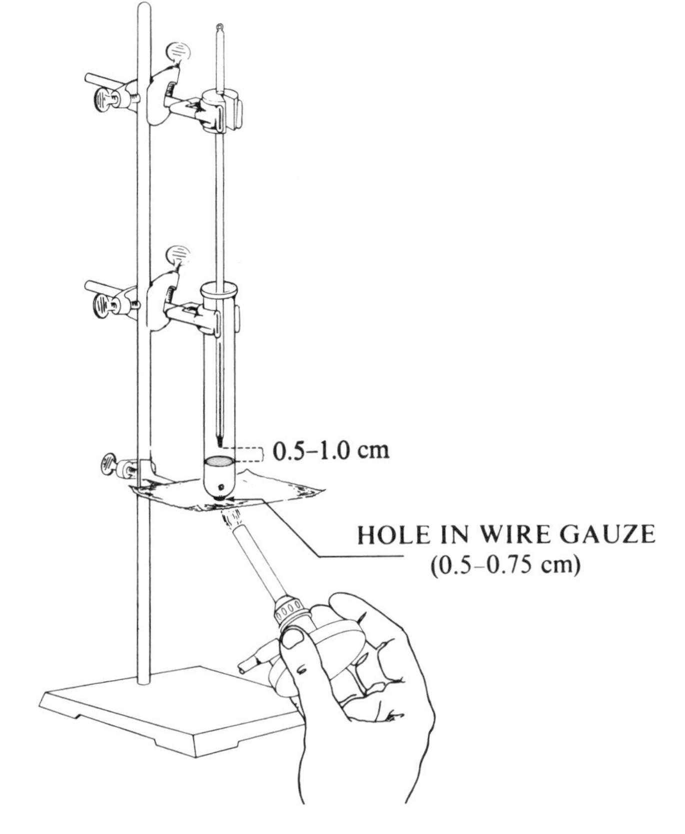

the test tube and add a small clay chip or a small Boileezer. Lower a 260° thermometer into the test tube until the mercury bulb is 0.5–1.0 cm above the surface of the liquid. Center the thermometer carefully to keep it clear of the inner wall of the test tube and clamp it securely in a cut-out cork stopper. Adjust your burner to a *very* small flame and manipulate it beneath the hole in the gauze until a *gentle* boiling of the liquid takes place. (Gentle boiling is essential for the success of this determination. Violent boiling will result in a rapid loss of the sample and possibly increase the danger of its becoming ignited by the burner flame.) The boiling point of the liquid is reached when the vapor condenses on the thermometer bulb and drips off in a series of steady drops and the mercury column of the thermometer no longer rises. Record the boiling point of your sample on the report form and extinguish your burner.

PART II PROPERTIES AND REACTIONS OF PHENOLS

Procedure

A. SOLUBILITY IN WATER AND ALKALINE SOLUTIONS

When the phenyl group is present as a substituent in an aliphatic compound (alcohol, aldehyde, ketone, acid, etc.), its effect on the solubility of the compound in water is approximately equivalent to that of a four-carbon alkyl group. Thus, benzyl alcohol has about the same solubility in water as *n*-amyl alcohol. This effect carries over to some extent to aromatic compounds as well. Benzoic acid is only slightly less soluble than *n*-valeric acid and phenol has approximately the same solubility as *n*-butyl alcohol.

Phenols are more acidic than the alcohols but less so than the carboxylic acids or carbonic acid. Phenols thus will form salts when treated with sodium hydroxide but not with sodium carbonate. This difference in acid strength serves to distinguish phenols from alcohols and from carboxylic acids.

$$NaOH + C_6H_5OH \longrightarrow C_6H_5O^-Na^+ + H_2O$$

Phenol Sodium phenoxide

Into three separate test tubes, place small samples (either 10 drops of liquid or a few small crystals (0.1 g) of a solid) of the following: phenol, *p*-chlorophenol, and 2-naphthol. **(CAUTION!** *Avoid contact with phenols, particularly low-molecular weight, more water-soluble, more volatile phenols. Phenol itself can cause very serious burns.*) Add 5 ml of water to each sample, mix, and observe. Next, place all test tubes in a water bath and warm to 65°. Test a drop of each solution on blue litmus paper. Remove all samples from the water bath and allow them to cool to room temperature. Observe and record your results. Save all solutions for procedure (C).

Repeat procedure (A) but, instead of water, use equivalent volumes of 5% sodium hydroxide solution. Mix by shaking each solution as before but do not heat. Again observe and record your results.

Repeat procedure (A) but, instead of water, use equivalent volumes of 5% sodium bicarbonate solution. Mix by shaking each solution as before but do not heat. Again observe and record your results.

B. REACTION WITH BROMINE

Bromine reacts very readily with phenols to introduce one or more bromine atoms in positions *ortho* or *para* to the phenolic hydroxyl group.

$$C_6H_5OH + 3\,Br_2 \rightarrow C_6H_2Br_3OH + 3\,HBr$$

Add bromine water dropwise to each of the solutions you prepared in procedure (A) until the color of bromine is no longer discharged. Shake well after each addition. Record your results.

C. FERRIC CHLORIDE TEST

Phenols and enols often react with ferric chloride to produce characteric colors. Not all phenols and enols give colors; therefore, a negative test is not necessarily significant.

To very dilute (0.1%) solutions of (1) phenol, (2) resorcinol, and (3) salicylic acid made by dissolving a few crystals of each in 5 ml of water, add a drop of 1% ferric chloride solution. Try the same tests on a sample of ethyl alcohol and also on a methanolic solution of ethyl acetoacetate. Record your results on the report form.

TEXT REFERENCE

Report: 10

Chapter Pages

Section Desk NAME

PROPERTIES AND REACTIONS OF THE ALCOHOLS

SOLUBILITY IN WATER AND IN PHOSPHORIC ACID.

Alcohol	Structure	Cold water	Hot water	85% H_3PO_4
n-Butyl alcohol	$CH_3(CH_2)_3OH$			
tert-Butyl alcohol	$CH_3{-}C(CH_3)_2{-}OH$			
Cyclohexanol	(cyclohexane ring)—OH			
Benzyl alcohol	(benzene ring)—CH_2—OH			

REACTION WITH SODIUM

Alcohol	Structure	Result
Methyl alcohol	CH_3—OH	
Isopropyl alcohol	CH_3—CH(CH_3)—OH	
n-Propyl alcohol	CH_3—CH_2—CH_2—OH	

REACTION WITH LUCAS REAGENT

Alcohol	Time for reaction	Description of reaction mixture
n-Butyl alcohol		
sec-Butyl alcohol		
tert-Butyl alcohol		

OXIDATION OF ALCOHOLS

Alcohol	Structure	Color changes observed	Color of litmus paper
Ethanol	CH_3—CH_2—OH		
Isopropyl alcohol	CH_3—CH—OH | CH_3		
tert-Butyl alcohol	CH_3 | CH_3—C—OH | CH_3		

Write balanced equations for the reactions (if any) that took place.

Ethanol:

Isopropyl Alcohol :

tert-Butyl Alcohol:

RESULTS OF THE IODOFORM REACTION

Compound	Structure	Result
Acetone	$CH_3-C(=O)-CH_3$	
Ethyl alcohol	CH_3-CH_2-OH	
Methyl alcohol	CH_3-OH	
Isopropyl alcohol	$CH_3-CH(OH)-CH_3$	
n-Propyl alcohol	$CH_3-CH_2-CH_2-OH$	

MELTING POINT OF IODOFORM

Experimental ________°C

Literature ________°C

Write balanced equations for the formation of iodoform from each compound above which gave a positive test.

DETERMINATION OF AN UNKNOWN

Unknown No. ________; Boiling point ________°C

Possible compounds ____

____

Results of distinguishing tests:

Lucas test____

Iodoform reaction____

Oxidation reaction____

Identity of unknown____

PROPERTIES AND REACTIONS OF PHENOLS

SOLUBILITY IN WATER, 5% SODIUM HYDROXIDE, AND 5% SODIUM BICARBONATE

Compound	Structure	Cold water	Hot water	5% NaOH	5% $NaHCO_3$
Phenol	OH (attached to benzene ring)				
p-Chlorophenol	Cl—(benzene ring)—OH				
2-Naphthol	(naphthalene ring)—OH				

REACTION WITH BROMINE

Compound	Result with bromine water
Phenol	
p-Chlorophenol	
2-Naphthol	

REACTION WITH FERRIC CHLORIDE

Compound	Structure	Result
Phenol		
Resorcinol		
Salicylic acid (consult text)		
Ethyl alcohol		
Ethyl acetoacetate (consult text)		

Questions and Exercises

1. Can you offer an explanation for the variation in reaction rate observed when primary, secondary, and tertiary alcohols react with the Lucas reagent?
2. An unknown alcohol gives a negative Lucas test and a positive iodoform test. Suggest a possible structure for the alcohol.
3. Suggest a simple method for the separation of phenol from a solution which contains both phenol and cyclohexanol.
4. An unknown reacts with sodium to form a sodium salt but further tests show the compound to be an aliphatic hydrocarbon. What structural unit is probably present in the hydrocarbon?
5. Theoretically, how much iodoform could be prepared from 25 ml of 95% ethanol (sp. gr. 0.79)?
6. What is a medicinal use for iodoform?
7. Beginning with benzene, show how you might prepare benzoic acid, C_6H_5—COOH, using the iodoform reaction in one step of the synthesis.
8. Only one aldehyde and one primary alcohol give a positive iodoform reaction. What compounds are these? Explain why they alone of their class give a positive reaction.
9. Systematic names are assigned to the compounds in Table 10.1. Can you assign another acceptable name to each?
10. Why do alcohols that show little tendency to dissolve in water appear to dissolve readily in 85% phosphoric acid? Would you describe the latter behavior as solubility or a chemical reaction? Explain.
11. Explain why ethyl acetoacetate was expected to react with ferric chloride in Part II-C of this experiment. (*Hint:* Consult the section describing "tautomerism" in your text.)
12. Draw the structures of 2-naphthol and benzhydrol (Exp. 11). Describe a simple chemical test for distinguishing between these two compounds.
13. * Both ir and nmr spectra of alcohols and phenols show similar absorption ranges for the hydroxyl group. In which regions of both types of spectra must we seek additional absorption evidence that will help us to distinguish a phenol from an aliphatic alcohol?
14. The aliphatic alcohols and ethers are isomeric; seven structures may be written for the molecular formula, $C_4H_{10}O$. If you were required to identify an unknown of this formula, tell exactly how you would go about it.
15. * If the nmr spectrum of the compound in Question 14 consisted of only two signals — one a triplet at δ 1.15 and the other a quadruplet at δ 3.48 — which of the seven compounds would be indicated?

Experiment 11

Time: 4 hours

Reactions of Aldehydes and Ketones

The aldehydes, $R{-}\overset{\overset{\displaystyle H}{|}}{C}{=}O$, and the ketones, $R{-}\overset{\overset{\displaystyle R}{|}}{C}{=}O$, both have the carbonyl group, $\diagdown\!\!\!\!\!\!\diagup C{=}O$, as a functional group and usually react with the same reagents to give similar products. In general, the principal differences in their reactions are in the relative rates of reaction. The aldehydes usually react more rapidly than do the ketones and give reactions that more nearly go to completion. Most of the reactions of both classes of compounds take place by addition across the carbon-oxygen double bond, frequently followed by loss of water. This experiment illustrates the preparation of an aldehyde and reactions of both aldehydes and ketones as useful chemical tests for distinguishing between the two classes.

Procedure

A. PREPARATION OF FORMALDEHYDE

A general method of preparation for both aldehydes and ketones is the copper-catalyzed dehydrogenation of the appropriate alcohol.

$$CH_3OH \xrightarrow[\text{heat}]{\text{Cu—CuO,}} H{-}\overset{\overset{\displaystyle H}{|}}{C}{=}O + H_2O$$

This procedure is especially useful as a simple laboratory preparation of formaldehyde from methyl alcohol.

Using a 10-inch length of No. 14 copper wire wind a spiral around a 6-mm glass tube (4–5 turns) making the wound portion about 0.5 inch in length and leaving about 5 inches of straight wire. Insert the straight end into a No. 7 cork. The cork serves the dual function of test tube stopper and insulating handle. If prepared properly, the copper spiral will almost reach the bottom of a 15-cm test tube when the cork is fitted loosely in the test tube mouth. Oxidize the spiral by heating it in the oxidizing portion of a burner flame (Fig. 11.1). Allow the spiral to cool and then examine it. Place 2 ml of

FIGURE 11.1 Preparation of the cupric oxide catalyst by heating a spiral of copper wire.

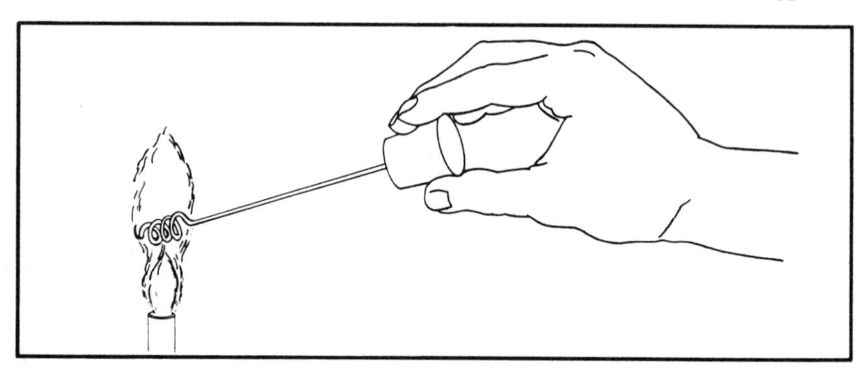

methyl alcohol and 5 ml of water in a 15-cm test tube. Again heat the copper spiral in the flame to red heat and immediately plunge it into the methyl alcohol solution. Allow the cork handle to stopper the test tube loosely. Cool the test tube in ice water. Withdraw and examine the copper spiral. Repeat the heating and quenching operation three or four times in the manner described above. Use the formaldehyde you have prepared in the tests of procedure (C).

B. PREPARATION OF ACETALDEHYDE

Acetaldehyde boils at 20° and is not easily stored at room temperatures in the usual reagent bottles. The trimer of acetaldehyde, paraldehyde,[1] on the other hand, boils at 125° and is a stable, easily stored liquid. This portion of the experiment illustrates the ready depolymerization of paraldehyde to the monomer, acetaldehyde.

Set up a distillation apparatus as illustrated in Figure 3.1, p. 23. Cool the receiver in an ice bath.

Pour 10 ml of paraldehyde into the distilling flask (using the long-stemmed funnel) and add to it 0.5 ml (10 drops) of a 1:1 concentrated sulfuric acid-water mixture. Heat the mixture gently over a low flame so that the temperature of the vapors which distill is maintained between 35–40°. Control the heating operation by holding the burner in your hand. Overheating not only causes

[1]**Note to Instructor.** In view of the fact that paraldehyde is a Schedule IV controlled substance and may be purchased only by a registrant under The Controlled Substances Act of 1970, some instructors may wish to furnish acetaldehyde for subsequent tests. Registration information may be obtained from the Drug Enforcement Administration, Department of Justice, P.O. Box 28083, Central Station, Washington, D.C. 20005.

carbonization, but also liberates sulfur dioxide which passes over into the distillate and catalyzes repolymerization of the product. Continue the heating operation until about 6 ml of liquid has been collected. Do not distill beyond this point. Stopper the flask loosely with a cork and keep it in the ice bath. Save the product for use in procedure (C). Do not try to keep the product for more than one laboratory period. Note the color and odor of your product.

C. TESTS FOR DISTINGUISHING BETWEEN ALDEHYDES AND KETONES

The chemical tests used to distinguish between aldehydes and ketones depend upon the difference in ease of oxidation of the two classes of compounds or upon differences in reactivity toward certain reagents. Although both aldehydes and ketones may be oxidized to acids, ketones are oxidized only under vigorous oxidizing conditions (i.e., stronger reagents and high temperatures). Under these conditions cleavage of the carbon chain occurs. Aldehydes, on the other hand, are readily oxidized under very mild conditions without cleavage of the carbon chain. The principal oxidizing agents used in the test reactions are the silver ion and the cupric ion. These ions are the active agents in two test solutions called, respectively, Tollens' Reagent and Benedict's Solution. The cupric ion is also the active agent in Fehling's solution. Another useful test reaction is based on a difference in reactivity of the two classes toward a test solution called *Schiff's Reagent.*

Tollens' Reagent is an ammoniacal solution of silver oxide. The reagent is reduced by aldehydes to metallic silver, and the aldehyde is oxidized to the corresponding acid. Ketones do not react with the reagent.

$$\left.\begin{array}{l} 2\,AgNO_3 + 2\,NaOH \rightarrow Ag_2O + H_2O + 2\,NaNO_3 \\ Ag_2O + 4\,NH_4OH \rightarrow 2\,Ag(NH_3)_2OH + 3\,H_2O \end{array}\right\} \text{Reactions in the preparation of Tollens' Reagent}$$

$$R{-}\overset{\displaystyle H}{\overset{|}{C}}{=}O + 2\,Ag(NH_3)_2OH \rightarrow R{-}\overset{\displaystyle O}{\overset{\|}{C}}{-}ONH_4 + 2\,Ag + NH_4OH$$

An aldehyde

To prepare Tollens' Reagent, clean four test tubes very carefully, using soap and water, and rinse them thoroughly with distilled water. To each test tube add 2 ml of a 5% solution of silver nitrate, add *one drop* of 10% sodium hydroxide solution, and mix the ingredients thoroughly by swirling. Dilute 7 ml of concentrated ammonium hydroxide (28% ammonia) to 100 ml with distilled water. Add the dilute ammonia solution dropwise to each test tube with constant swirling or shaking, until the brown-gray precipitate of silver oxide *just dissolves.* The test will fail if a large excess of ammonia is added.

To one of four test tubes containing Tollens' reagent add a few drops of the formaldehyde solution prepared in procedure (A). Mix by shaking and warm the tube gently on the water bath. If the test tube is very clean and the reactants are not too concentrated, a silver mirror will form on the inner surface of the test tube. If the test tube is not clean, a black precipitate of finely divided silver will form. Repeat the test with *acetaldehyde, benzaldehyde,* and *acetone.* Record your results. (The silver mirror is easily removed from a glass surface with dilute nitric acid.)

CAUTION! *Discard any unused Tollens' Reagent after you have completed this portion of the experiment.* Explosive silver nitrogen compounds form on prolonged standing.

Fehling's Solution[1] is a solution of cupric hydroxide (stabilized as a tartrate complex). *Benedict's Solution*[1] differs from Fehling's solution in that the cupric ion is stabilized as a citrate rather than as a tartrate. Both reagents are blue in color but are reduced by aldehydes to produce red cuprous oxide. In rare cases metallic copper is obtained. Neither reagent is affected by ketones. For simplicity, the copper complex may be represented as CuO (cupric oxide) in writing equations.

$$\underset{\text{An aldehyde}}{R{-}\overset{\overset{\displaystyle H}{|}}{C}{=}O} + 2\,CuO + Na^{+}OH^{-} \rightarrow \underset{\text{Sodium salt of an acid}}{R{-}\overset{\overset{\displaystyle O}{\|}}{C}{-}O^{-}Na^{+}} + Cu_2O + H_2O$$

To each of four test tubes add either 5 ml of Benedict's solution or 5 ml of freshly prepared Fehling's solution (made by mixing 10 ml of Fehling's solution A and 10 ml of Fehling's solution B). To one tube add 1–2 ml of the formaldehyde solution, to the second add 1–2 ml of acetaldehyde, to the third add 1–2 ml of benzaldehyde, and to the fourth add 1–2 ml of acetone. Shake the test tubes *vigorously*, then place them in a boiling water bath 10–15 minutes. In a positive test the solution first turns a pale green, then a reddish precipitate of cuprous oxide forms. If no precipitate forms, add a few more drops of the carbonyl compound and continue heating as before. Record your observations.

Schiff's Fuchsin-Aldehyde Reagent[1] is a sulfur dioxide-decolorized solution of the pink dye, fuchsin. Schiff's reagent reacts with aldehydes to produce a highly colored dye. The color produced usually is pink but may have a definite purple or blue cast. Ketones do not give a positive test when pure, but the Schiff's test is very sensitive and will be positive even if only a trace of aldehyde is present.

To each of five test tubes add 1 ml of Schiff's reagent (Note 1). To the first test tube add a few drops of the formaldehyde solution prepared in procedure (A), to the second a drop of acetaldehyde in 5 ml of water, to the third two drops of acetone in 5 ml of water, and to the fourth one drop of benzaldehyde in 5 ml of water. As a blank, add 1 ml of dilute sodium hydroxide to the

[1]See appendix.

fifth test tube. The latter treatment regenerates the original pink dye. Observe and record all color changes.

D. HEXAMETHYLENETETRAMINE (UROTROPINE, HEXAMINE)

Formaldehyde is unique in many respects when compared with the other aldehydes. It reacts with ammonia to yield the urinary antiseptic called **urotropine** or **hexamethylenetetramine.**

$$6\,H{-}\overset{\displaystyle H}{\overset{|}{C}}{=}O + 4\,NH_3 \rightarrow \underset{\text{Hexa-methylenetetramine}}{(CH_2)_6N_4} + 6\,H_2O$$

To 1 ml of formalin (35–40% formaldehyde) in an evaporating dish add 3 ml of concentrated ammonium hydroxide. Evaporate the mixture to dryness on the water bath. The whitish residue which remains is hexamethylenetetramine.

E. BISULFITE ADDITION PRODUCT

Aldehydes and some ketones (methyl ketones and cyclic ketones) react with sodium bisulfite to yield bisulfite addition products.

$$R{-}\overset{\displaystyle H}{\overset{|}{C}}{=}O + NaHSO_3 \rightarrow R{-}\underset{\underset{\displaystyle H}{|}}{\overset{\overset{\displaystyle OH}{|}}{C}}{-}SO_3^-Na^+$$

Place 5 ml of a *saturated* solution of sodium bisulfite in a 50-ml Erlenmeyer flask and cool the solution in an ice bath. Add 2.5 ml of acetone *dropwise* with swirling to mix the reactants. Allow the mixture to chill for about five minutes, then add 10 ml of ethyl alcohol. Stir the mixture and filter the crystals on the Hirsch funnel. Repeat the above procedure using cyclopentanone and diethyl ketone. Record your results.

F. THE REDUCTION OF CARBONYL COMPOUNDS WITH SODIUM BOROHYDRIDE

Sodium (or potassium) borohydride is a highly selective reagent that reduces aldehydes or ketones to the corresponding alcohols but normally does not reduce nitro, nitrile, olefinic, amide, carboxylic acid, or ester functional groups. The reagent is less reactive (i.e., more selective) than lithium aluminum hydride. Moreover, it may be used in aqueous or alcoholic solutions. Lithium aluminum hydride, on the other hand, reacts violently with such hydroxylic compounds and must be used in inert solvents such as ether, tetrahydrofuran, etc. In this experiment a simple example of the use of sodium borohydride is illustrated by the reduction of benzophenone (diphenyl ketone) to benzhydrol (diphenylcarbinol).

$$4\,(C_6H_5)_2C{=}O + NaBH_4 \rightarrow \left(H{-}\underset{C_6H_5}{\overset{C_6H_5}{C}}{-}O\right)_4 BNa$$

Benzophenone

$$\left(H{-}\underset{C_6H_5}{\overset{C_6H_5}{C}}{-}O\right)_4 BNa + HCl + 3\,H_2O \rightarrow 4\,(C_6H_5)_2CH{-}OH + H_3BO_3 + NaCl$$

Benzhydrol

Procedure

Dissolve 4.5 g (0.025 mole) of benzophenone in 30 ml of methanol in a 150-ml beaker. In a smaller beaker prepare a solution of 1.0 g of sodium borohydride in 15 ml of cold water. Add the aqueous sodium borohydride solution in small portions and with stirring to the benzophenone solution at such a rate that the temperature does not exceed 45°. The reaction is exothermic and the rate of addition, therefore, should not be too rapid. After all the sodium borohydride solution has been added, continue to stir the reaction mixture for approximately 15 minutes or until the diphenylcarbinol begins to precipitate. Continue to stir until a heavy slurry of crystals form. Decompose the excess sodium borohydride by adding the crystalline slurry slowly and with stirring to a mixture of 100 g of crushed ice and water and 10 ml of concentrated hydrochloric acid in a 400-ml beaker (Note 2). Collect the diphenylcarbinol on the Büchner funnel, wash the crystal cake twice with 50-ml portions of water and dry. The product is of a high degree of purity but may be recrystallized from aqueous ethanol. Dry the product, weigh it, and determine its melting point. Yield about 4.4 g (95%). Place your product in a properly labeled sample bottle and turn it in with your report.

Note 1. Because of the sensitivity of the Schiff reagent to traces of aldehyde, the laboratory supply may be spoiled if each student pours the reagent directly from the stock bottle into his aldehyde samples. Thus, the student should obtain the necessary amount of reagent in a small flask and conduct all tests at his bench.

Note 2. The decomposition of sodium borohydride is a vigorous reaction and hydrogen is liberated rapidly. Unless a large capacity beaker is used the froth which results from the decomposition cannot be confined.

TEXT REFERENCE

Report: 11

Chapter Pages

Section Desk NAME

REACTIONS OF ALDEHYDES AND KETONES

In the table below record after each substance tested the results obtained with each reagent used.

Compound	Tollens' reagent	Fehling's solution	Schiff's	$NaHSO_3$
Formaldehyde				
Acetaldehyde				
Acetone				
Benzaldehyde				
Cyclopentanone				
Diethyl ketone				

BENZHYDROL (DIPHENYLCARBINOL)

Reaction equation $(C_6H_5)_2CO \xrightarrow[\text{2. HCl, 3 } H_2O]{\text{1. } NaBH_4} (C_6H_5)_2CHOH$

Benzophenone Benzhydrol

	Benzophenone	Benzhydrol
Quantities	4.5 g	________
Mol. Wt.	________	________
Moles	________	________

Theoretical yield ________ g

Actual yield ________ g

Percentage yield ________

mp ________°C

Questions and Exercises

1. Interpret the color changes in the copper spiral used in procedure (A) of the experiment when the spiral was heated in the oxidizing flame of the burner; when plunged into the methanol solution.
2. Suggest a simple test that will serve to distinguish between (a) benzaldehyde, (b) valeraldehyde, and (c) acetophenone.
3. Can you suggest a reason why diethyl ketone fails to form an addition product with sodium bisulfite while cyclopentanone gives an addition product?
4. Draw the structure of hexamethylenetetramine. (Clue: Every CH_2 group has only N's as neighbors; every N has only CH_2 groups as neighbors.)
5. Write equations for the reduction of the carbonyl groups in the following compounds with the appropriate metal hydride: (a) cyclohexanone, (b) *n*-valeric acid, (c) carbon dioxide, and (d) benzamide.
6. Write the equation for an alternative synthesis of benzhydrol from a carbonyl compound in which reduction is not involved.

* 7. What singular structural feature of both aldehydes and ketones is most readily observed in the ir spectra of these classes of compounds?

* 8. The formula C_4H_8O may be that of either a ketone or two aldehydes. How do the nmr spectra of these three compounds differ with respect to: (1) number of signals; (2) splitting of signals; (3) shift regions?

9. Plan a separation scheme (other than fractional distillation) by making a flow chart outlining the steps required to separate a mixture containing acetone, 2-propanol, and propionic acid.

Experiment 12

Time: 3½ hours

Identification of an Unknown Carbonyl Compound

The formation of a crystalline derivative that is easy to purify and that possesses a sharp melting point is an excellent means of identifying an unknown organic compound. Such derivatives are especially helpful if the unknown is a liquid. The carbonyl compounds (aldehydes and ketones) form crystalline condensation products with a number of ammonia derivatives, but for identification purposes the hydrazones and the semicarbazones are most frequently prepared. Aldehydes and ketones react with phenylhydrazine to produce derivatives called **phenylhydrazones.** Derivatives formed from semicarbazide are called **semicarbazones.**

$$>C=O + H_2N-NH-C_6H_5 \rightarrow >C=N-NH-C_6H_5 + H_2O$$

Phenylhydrazine A phenylhydrazone

$$>C=O + H_2N-NH-C(=O)-NH_2 \rightarrow >C=N-NH-C(=O)-NH_2 + H_2O$$

Semicarbazide A semicarbazone

Often in qualitative work the prepared derivative of an unknown carbonyl compound is the **2,4-dinitrophenylhydrazone.**

$$>C=O + H_2N-NH-C_6H_3(NO_2)_2 \rightarrow >C=N-NH-C_6H_3(NO_2)_2 + H_2O$$

2,4-Dinitrophenylhydrazine A 2,4-dinitrophenylhydrazone

The substituted phenylhydrazone usually has a melting point higher than that of the unsubstituted phenylhydrazone. Moreover, certain carbonyl compounds form noncrystalline derivatives with phenylhydrazine. When this is the case, the 2,4-dinitrophenylhydrazone or the semicarbazone is prepared. The melting point of a phenylhydrazone, a 2,4-dinitrophenylhydrazone, or a semicarbazone, when compared with those recorded for carbonyl derivatives in tables and handbooks, often will serve to identify an unknown aldehyde or ketone. However, inasmuch as two or more different aldehydes and ketones may react with a reagent to give derivatives which melt at or near the same temperature, it frequently becomes necessary to prepare more than one derivative of a compound in order to identify it. It would be most unusual, however, for a compound to form two or more different derivatives with melting points near those of the corresponding derivatives of another substance. In this experiment the phenylhydrazones, the 2,4-dinitrophenylhydrazones, and the semicarbazones of several carbonyl compounds are prepared. You then will be required to identify an unknown carbonyl compound.

Procedure

A. PREPARATION OF PHENYLHYDRAZONES

Dissolve 2 g of phenylhydrazine hydrochloride, $C_6H_5NHNH_2 \cdot HCl$, in 18 ml of warm distilled water. **CAUTION!** *Phenylhydrazine is very toxic. Avoid contact with skin.* Add 3 g of crystalline sodium acetate, $CH_3COONa \cdot 3\ H_2O$, and one drop of glacial acetic acid. If the solution is turbid, add a micro-spatula measure of charcoal, shake the solution vigorously, and filter. The solution deteriorates rapidly. Prepare just before using.

Divide the phenylhydrazine reagent into two equal portions in two test tubes. Save one portion for identification of your unknown, procedure (D). To the other portion add 0.5 ml (10 drops) of benzaldehyde, C_6H_5CHO. Stopper the tube and shake vigorously until the product crystallizes. Filter the crystals with suction using the small Hirsch funnel and wash them thoroughly with water. Recrystallize the product from ethyl alcohol using the following procedure.

Transfer the crystals to a 50-ml Erlenmeyer flask, and add dropwise 1–2 ml of ethyl alcohol. Heat the mixture on the steam bath. If the product dissolves very readily in the initial amount of alcohol, add water to the alcoholic solution *dropwise* until a *faint* turbidity persists in the solution while it is hot. Add ethyl alcohol dropwise to the hot solution until it again becomes clear, then add one or two drops of alcohol in excess. Proceed as in the next paragraph. If the product does not dissolve completely in the initial volume of alcohol after boiling for a few minutes, add more ethyl alcohol dropwise and with heating, repeating the process until all the solid dissolves. Use no more alcohol than necessary to effect solution of your product. **CAUTION!** *Ethyl alcohol is very flammable.*

If the solution is dark or highly colored, add a small amount of charcoal and filter the solution. Cool the filtrate in the ice bath until the product crystallizes, filter the crystals with suction on the Hirsch funnel, wash them with a small volume (1 ml) of *cold* ethyl alcohol, and dry them as quickly as possible on a piece of filter paper. Transfer a small amount of the product to a piece of clay plate or to a piece of filter paper on a watch glass, complete the drying by crushing the product on the plate (or paper) and determine the melting point.

B. PREPARATION OF SEMICARBAZONES

Dissolve 2 g of semicarbazide hydrochloride, $H_2N—NHCONH_2 \cdot HCl$, in 20 ml of distilled water. Add 3 g of crystalline sodium acetate. Mix and divide the reagent into two equal portions. Save one portion as before. To the other portion add 0.5 ml (10 drops) of methyl ethyl ketone. Stopper the test tube with a cork and shake vigorously. Filter the crystals using the Hirsch funnel and recrystallize from ethyl alcohol as described in procedure (A). When the product is dry, determine its melting point.

C. PREPARATION OF 2,4-DINITROPHENYLHYDRAZONES

Dissolve 0.5 g of 2,4-dinitrophenylhydrazine in 3 ml of concentrated sulfuric acid. Add this solution slowly, with stirring, to 5 ml water and 15 ml of 95% ethyl alcohol. Mix thoroughly. Divide the mixture into *three* equal portions. Save one portion as before. To one portion add 0.5 ml (10 drops) of methyl ethyl ketone with stirring. The 2,4-dinitrophenylhydrazone usually forms immediately. If no precipitate results, set the reaction mixture aside for 15 minutes. Shake occasionally and scratch the wall of the test tube with a glass stirring rod to help induce crystallization. If the crystalline slurry which forms should be too heavy to filter conveniently, add 10–15 ml of water. Prepare the 2,4-dinitrophenylhydrazone of benzaldehyde using the second portion of your reagent and following the procedure just described. Separate and purify both derivatives according to the procedure discribed under procedure (A) and determine their melting points.

D. IDENTIFICATION OF AN UNKNOWN CARBONYL COMPOUND

Obtain 5 ml of an unknown carbonyl compound from your laboratory instructor. Prepare the phenylhydrazone of your unknown, using the second portion of the phenylhydrazine reagent left from procedure (A). Recrystallize the phenylhydrazone and determine its melting point. Prepare the semicarbazone of your unknown, using the second portion of the reagent left from procedure (B). From Table 12.1 determine the identify of the unknown.

If the unknown carbonyl compound issued to you forms a noncrystalline phenylhydrazone, it may be necessary for you to prepare a 2,4-dinitrophenylhydrazone as well as a semicarbazone. Should this be the case, use the remaining portion of the reagent saved from procedure (C). The semicarbazones usually are higher melting derivatives than are the phenylhydrazones.

TABLE 12.1
MELTING POINTS OF SOME ALDEHYDE AND KETONE DERIVATIVES (°C)

Compound and mp	Structure	Phenylhydrazone	Semicarbazone	2,4-Dinitrophenylhydrazone
Acetone (56)	$CH_3-C(=O)-CH_3$	42	187	128
n-Butyraldehyde (74)	$CH_3CH_2CH_2CHO$	oil	104	123
Methyl ethyl ketone (80)	$CH_3-C(=O)-C_2H_5$	oil	146	116
Diethyl ketone (102)	$C_2H_5-C(=O)-C_2H_5$	oil	139	156
Methyl *tert*-butyl ketone (106)	$(CH_3)_3C-C(=O)-CH_3$	oil	157	125
Cyclopentanone (131)	(cyclopentane ring)=O	50	205	142
Cyclohexanone (155)	(cyclohexane ring)=O	77	166	162
Acetophenone (202)	$C_6H_5-C(=O)-CH_3$	105	199	239
Benzaldehyde (179)	C_6H_5-CHO	158	222	237
Furfural (161)	(furan ring)—CHO	97	202	229

TEXT REFERENCE

Report: 12

Chapter Pages

Section Desk NAME

DERIVATIVES OF ALDEHYDES AND KETONES

Compound	Phenylhydrazone	Semicarbazone	2,4-Dinitrophenyl-hydrazone
Benzaldehyde	mp ______°C (experimental) mp ______°C (Table 12.1)		mp ______°C (experimental) mp ______°C (Table 12.1)
Methyl ethyl ketone	(noncrystalline)	mp ______°C (experimental) mp ______°C (Table 12.1)	mp ______°C (experimental) mp ______°C (Table 12.1)

IDENTIFICATION OF AN UNKNOWN CARBONYL

Number of unknown ________

Derivatives formed and melting points: ______________________

____

Identity of unknown carbonyl compound: ______________________

Experiment 13

Time: 3 hours

Reactions of the Amines

The amines are the principal organic bases. Structurally, they are related to ammonia and have the general formulas RNH_2, R_2NH, and R_3N in which one, two, or all three of the hydrogen atoms of the ammonia molecule have been replaced by alkyl or aryl groups. Amines are classified as primary, secondary, or tertiary according to the number of hydrogen atoms of ammonia that have been replaced.

The amines are basic compounds because only three of the five electrons in the valence shell of the nitrogen atom are used in covalent bonding. An amino nitrogen with two unshared electrons can function as an electron-pair donor, or base. A water-soluble amine, like ammonia, may combine reversibly with water to produce a basic, substituted ammonium hydroxide according to the following equation.

$$R{-}\underset{\substack{|\\H}}{\overset{\substack{H\\/}}{N}}: + H{-}\underset{\substack{|\\H}}{\ddot{O}}: \rightleftharpoons \left[R{-}\underset{\substack{|\\H}}{\overset{\substack{H\\|}}{N}}:H\right]^+ :\ddot{\underset{..}{O}}H^-$$

Moreover, structures deficient by an electron pair (acids) may combine with an amine to produce salts.

$$R{-}\underset{\substack{|\\H}}{\overset{\substack{H\\|}}{N}}: + H^+Cl^- \rightarrow \left[R{-}\underset{\substack{|\\H}}{\overset{\substack{H\\|}}{N}}:H\right]^+ Cl^-$$

The following series of experiments illustrate the properties of the amines.

Procedure

A. BASIC PROPERTIES OF THE AMINES

Into four separate test tubes place small samples (about 10 drops each) of the following amines: aniline, benzylamine, diethylamine, and pyridine. To each test tube add 5 ml of water, stopper with a clean cork, and shake to mix. Are all amines completely soluble in water to give a homogeneous solution? Record your observations. Test each solution with pink litmus paper.

Are these solutions basic? Which of the solutions tested appears to be the strongest base?

Now add concentrated hydrochloric acid (use a dropper) to the test tubes containing the amine samples until each solution is distinctly acid to litmus. Have those amines that were not completely soluble in water now gone into solution? Write equations on your report form for the reactions that took place when HCl was added to each test tube. Mark for identification and save all solutions of amine hydrochlorides for procedure (C).

B. SALT FORMATION

Place 10 ml of *anhydrous* ether in a clean, dry test tube and add 10 drops of aniline. Take the ether solution of aniline to the fume hood and bubble *dry* hydrogen chloride from the HCl generator (Fig. 13.1) through the solution for a few minutes until precipitation appears to be complete. Write the equation for this reaction on your report form. Collect the white precipitate by suction filtration on a small Hirsch funnel. Save a small amount for a melting point determination and transfer the balance to a clean dry test tube. Add 5 ml of water. Result? Now test the aqueous solution with blue litmus. Also test the solution with a drop of silver nitrate solution. Now make the solution strongly basic with sodium hydroxide (use litmus). Shake and observe. Compare the appearance of the basic solution with that produced in the first part of procedure (A). Write equations for all reactions that have taken place.

FIGURE 13.1 Hydrogen chloride generator.

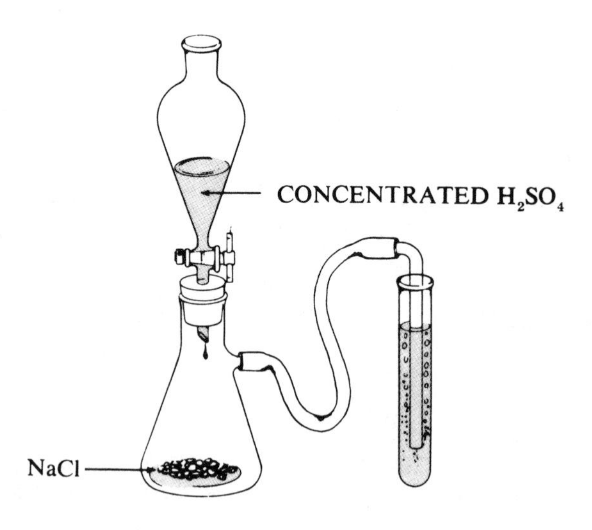

C. AMINES WITH NITROUS ACID

Nitrous acid is an unstable substance prepared in solution only when needed by reacting a mineral acid with sodium nitrite.

$$Na^+NO_2^- + H^+Cl^- \rightarrow Na^+Cl^- + HNO_2$$

When treated with nitrous acid a primary amine yields an alcohol as one product:

$$R{-}NH_2 + HNO_2 \rightarrow R{-}OH + H_2O + N_2$$

The reaction is carried out by treating an acid solution of the amine (an amine salt solution) with an aqueous solution of sodium nitrite.

Secondary amines react with nitrous acid to produce neutral N-nitroso compounds. These are yellow oils that may be separated from a solution containing other amines. (**CAUTION!** *N-Nitrosoamines are carcinogenic. Do not treat the hydrochloride of diethylamine with sodium nitrite solution.*)

$$R_2N{-}H + H{-}O{-}N{=}O \rightarrow R_2N{-}N{=}O + H_2O$$

An N-nitroso compound

Tertiary amines may simply dissolve in nitrous acid to form nitrite salts or undergo a rather complex degradation. The reaction is of little or no practical utility.

The reaction of aliphatic amines with nitrous acid has little value except as a diagnostic or as a separatory procedure. In either case, the primary amine is destroyed. On the other hand, the reaction of nitrous acid with aromatic primary amines produces intermediates known as diazonium salts.

$$C_6H_5{-}NH_2 + 2\,HCl + NaNO_2 \xrightarrow{0\text{–}5^\circ C} C_6H_5{-}N_2^+Cl^- + NaCl + H_2O$$

Benzenediazonium chloride

A large number of very useful products, difficult, if not impossible to arrive at by any other route, are synthesized by way of a diazonium salt. Many of our most useful dyes are azo dyes prepared in this way. [See Experiment 30, procedure (A)].

Place the test tubes containing only the hydrochloride solutions of aniline, benzylamine, and pyridine prepared in procedure (A) in a beaker of ice water. When the temperature has fallen to about 5° add to each test tube about 2 ml of a 5% solution of sodium nitrite. Observe closely and note if there is gas evolution from any of the test tubes. Remove the test tube containing the aniline sample and transfer it to a beaker of warm water (40°). Observe closely. Is there now any evolution of gas? Describe any change in appearance of the solution. Explain and write equations for the reactions that have taken place.

D. SOLID DERIVATIVES OF AMINES

A convenient method for the identification of unknown amines is to convert them into sharp-melting crystalline derivatives. Most primary and secondary amines react with acetic anhydride to form substituted acetamides. This reaction is illustrated in the preparation of acetanilide (Experiment 26).

$$(CH_3CO)_2O + RNH_2 \rightarrow CH_3CONHR + CH_3COOH$$

Acetic anhydride — N-substituted acetamide

Primary and secondary amines also form N-substituted benzamides but in this case the more reactive benzoyl chloride is used rather than the acid anhydride.

$$C_6H_5-C(=O)-Cl + H_2N-R \rightarrow C_6H_5-C(=O)-N(H)-R$$

Benzoyl chloride — N-substituted benzamide

(a) Place 5 drops of benzylamine in a test tube and add 10 drops of water. Add to the aqueous solution of the amine 5 drops of benzoyl chloride and 2 ml of 10% sodium hydroxide. (**CAUTION!** *Benzoyl chloride has a strong irritating vapor.*) Stir with a stirring rod. The N-substituted benzamide should form almost immediately. Break up any lumps that form with your stirring rod, add 5 ml of water and filter. Wash with an additional 10 ml of water and recrystallize from 95% ethanol, dry, and determine the melting point of your derivative.

Most primary and secondary amines react readily with phenylisocyanate and phenyl isothiocyanate to give substituted ureas and thioureas respectively.

$$R{-}NH_2 + C_6H_5{-}N{=}C{=}O \rightarrow C_6H_5{-}\underset{H}{N}{-}\overset{O}{\overset{\|}{C}}{-}\underset{H}{N}{-}R$$

Phenyl isocyanate N-substituted urea

$$R{-}NH_2 + C_6H_5{-}N{=}C{=}S \rightarrow C_6H_5{-}\underset{H}{N}{-}\overset{S}{\overset{\|}{C}}{-}\underset{H}{N}{-}R$$

Phenyl isothiocyanate

The phenyl isothioureas are especially easy to prepare because phenyl isothiocyanate is not sensitive to water as is phenyl isocyanate. The latter combines with water to give an unstable carbamic acid that decomposes to aniline and carbon dioxide.

(b) Place 5 drops of benzylamine in a test tube and add 5 drops of phenyl isothiocyanate. Mix with a stirring rod. A reaction should take place almost immediately to produce a solid crystalline cake. Add 10 ml of water to the test tube, break up the crystalline material, and collect it by suction filtration on your Hirsh funnel. Wash the crystalline material once with 5 ml of 50% ethanol and once with 5 ml of *n*-hexane or Skellysolve to remove unreacted materials. The derivative as collected is of a high state of purity but may be recrystallized from 95% ethanol. Determine and record the melting point.

TEXT REFERENCE

Report: 13

Chapter Pages

Section Desk NAME

A. BASIC PROPERTIES OF AMINES

Amine	Solubility in water	Action on pink litmus paper	Solubility in hydrochloric acid
Aniline			
Benzylamine			
Diethylamine			
Pyridine			

Strongest basic solution________________.

Reactions of amines with HCl:

B. SALT FORMATION

Aniline + dry HCl → Melting point of product________

Result upon adding water________________

Equation:

Action of aqueous solution on blue litmus paper________

Result upon adding a drop of $AgNO_3$________

Equation:

Result upon making solution basic______________________________

Equation:

Melting point of aniline hydrochloride__________

C. **AMINES WITH NITROUS ACID**

Complete the following reactions:

Aniline + $HNO_2 \xrightarrow{0\text{–}5^\circ}$

Aniline + $HNO_2 \xrightarrow{40^\circ}$

Explanation of color change__________________________________

Equation:

Benzylamine + $HNO_2 \xrightarrow{0\text{–}5^\circ}$

Pyridine + $HNO_2 \xrightarrow{0\text{–}5^\circ}$

D. DERIVATIVES OF THE AMINES

Equation for the reaction leading to a derivative of benzylamine with benzoyl chloride:

(a)

Melting point of substituted benzamide ____________ ____________
Experimental Literature

Equation for the reaction leading to a derivative of benzylamine with phenyl isothiocyanate:

(b)

Melting point of phenyl thiourea ____________ ____________
Experimental Literature

Questions and Exercises

1. Outline a scheme which will lead to the isolation of a water insoluble amine when it is present in a mixture that also includes benzoic acid and anisole.

2. Considering the ring-activating influence of the amino group of aniline, suggest an easily-prepared derivative of aniline in which only the hydrogen atoms of the ring are substituted.

3. The reaction of benzoyl chloride with primary and secondary amines in the presence of sodium hydroxide is referred to in textbooks as the Schotten-Baumann reaction. The concentration of hydroxide is a critical factor. A high concentration of sodium hydroxide dissolves the benzoyl derivatives of primary amines and is to be avoided. On the other hand, if the reaction medium were not slightly basic, a by-product melting at 122° might coprecipitate with the substituted benzamide. What might this compound be?

4. Suggest a reason why the benzenesulfonamides, $C_6H_5{-}SO_2{-}NH{-}R$, of primary amines are far more soluble in aqueous alkali than the corresponding benzamides.

* 5. The N-H stretching frequency of the primary amino group usually appears as a doublet at 3200–3500 cm^{-1}. If an ir spectrum shows this absorption along with a strong absorption at or near 1700 cm^{-1}, what conclusions should you draw?

* 6. An organic compound, $C_9H_{13}N$, gave an nmr spectrum showing a 5-proton singlet at δ 7.34. With only this much information, how many isomeric structures can you draw?

7. If the amine in Question 6 failed to react with nitrous acid, how many structures would be represented by $C_9H_{13}N$?

8. Aniline and benzylamine both boil within the same temperature range (184–185°). How could one be distinguished from the other by a simple chemical test?

9. Consult a table of derivatives for primary amines and tell why the phenylthiourea derivatives would be of no help in identifying either compound in Question 8. What would be a good derivative in this case?

Experiment 14

Time: Variable

The Identification of an Unknown Organic Compound: Infrared and Nuclear Magnetic Resonance Spectroscopy

No other exercise in the organic laboratory program will test your knowledge of organic chemistry more completely or be more instructive than one requiring the identification of an unknown organic compound.

For many years the identification of unknown organic compounds as a laboratory exercise was deferred until students had reached a level of attainment and expertise which enabled them to carry out many chemical test reactions. However, the introduction of instrumental methods has greatly simplified the identification of an unknown organic substance. Many functional groups and other structural features of compounds now are readily identified by spectroscopic methods, and much of what at one time rested entirely upon the judgment of the student is accomplished by the infrared and the nuclear magnetic resonance spectrometers. These two instruments are of special value to the organic chemist in any identification scheme, but confirmation of the identity of an unknown should always be established by some supplementary chemical procedures.

By this time your laboratory program has included sufficient exercises dealing with the properties of the various classes of compounds to enable you to classify and to identify any of the common organic substances listed in Table 14.1. In this experiment you will be issued one of the compounds listed in the table and you will be required to identify it. The ir and nmr spectra for each of the compounds in the table are identified by letter only and appear in the appendix. Your instructor will tell you which spectra are those of your unknown[1]. Study the spectra carefully and, with the help of Table 14.2 or one that appears in your text, try to derive as much structural information as you can that will enable you to place your unknown in its proper category—that is, among the acids, aldehydes, amines, alcohols, etc. From spectral studies alone it is possible to establish the identity of most of the compounds listed, but the practice of a little "wet" chemistry will help you make a positive identification. The next order of business, therefore, will be the accurate determination of a boiling point or a melting point of the compound. You should then test the solubility behavior of the unknown to help further classify it as an acid, an amine, or as one of the "neutral" compounds—aldehydes, ketones, esters, amides, etc. Finally, you will

[1] **Note to instructor.** A key to the identity of the compound corresponding to the spectra in the appendix should be known to the instructor only and may be obtained from the publisher.

want to carry out one or more of the simple chemical tests described in previous experiments that deal with the chemical properties of the different classes of compounds.

Record on the report form all the information from the above studies that has led to the identity of your unknown compound.

TABLE 14.1
COMPOUNDS TO BE ISSUED AS UNKNOWNS

Compound	Structure	bp(°C)	mp(°C)
4-Chlorobenzaldehyde	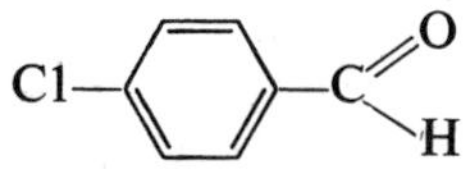 $Cl-C_6H_4-CHO$		48
Benzhydrol	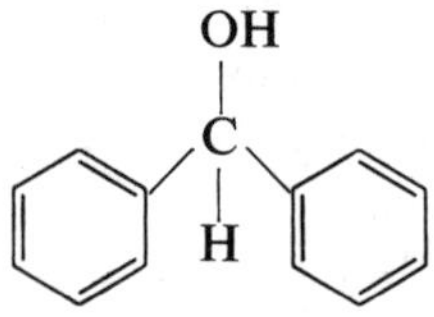 $(C_6H_5)_2CH(OH)$		69
4-Chloroaniline	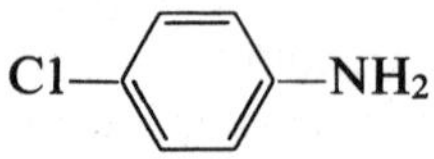 $Cl-C_6H_4-NH_2$		70
Ethyl acetate	$CH_3-C(=O)OCH_2CH_3$	77	
2-Butanone	$CH_3-C(=O)CH_2CH_3$	80	
2-Propanol	$CH_3-CH(OH)-CH_3$	82	
2-Methyl-2-propanol	$CH_3-C(CH_3)(OH)-CH_3$	83	
Propanol	$CH_3CH_2CH_2OH$	98	
3-Pentanone	$CH_3CH_2-C(=O)-CH_2CH_3$	102	

TABLE 14.1 (*Continued*)
COMPOUNDS TO BE ISSUED AS UNKNOWNS

Compound	Structure	bp(°C)	mp(°C)
1, 2-Dihydroxybenzene (Catechol)	$C_6H_4(OH)_2$ (1,2-)		106
Benzoic acid	C_6H_5—C(=O)OH		122
o-Benzoylbenzoic acid	C_6H_4(COOH)(C(=O)C_6H_5)		127
Benzamide	C_6H_5—C(=O)NH_2		128
Cyclopentanone	cyclopentane=O	131	
trans-Cinnamic acid	C_6H_5—CH=CH—C(=O)OH		133
Ethylbenzene	C_6H_5—CH_2CH_3	136	
Anisole	C_6H_5—OCH_3	153	
Benzaldehyde	C_6H_5—C(=O)H	179	
Benzylamine	C_6H_5—CH_2NH_2	184	
N-Methylaniline	C_6H_5—N(H)CH_3	196	

TABLE 14.1 (*Continued*)
COMPOUNDS TO BE ISSUED AS UNKNOWNS

Compound	Structure	bp(°C)	mp(°C)
Acetophenone	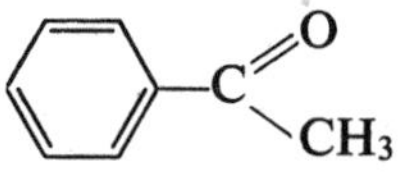 C_6H_5—C(=O)—CH_3	200	
m-Cresol	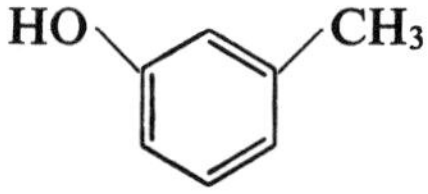 HO—C_6H_4—CH_3 (meta)	203	
Benzyl alcohol	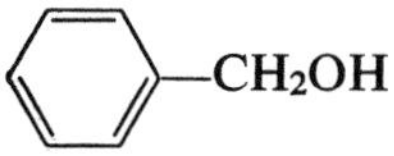 C_6H_5—CH_2OH	204	
4-Ethoxyaniline (*p*-Phenetidine)	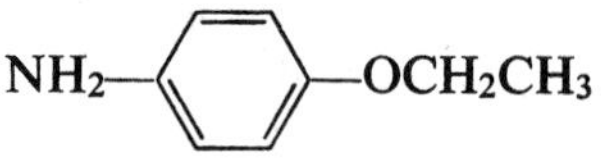 NH_2—C_6H_4—OCH_2CH_3	250	

TABLE 14.2
A SIMPLIFIED CORRELATION TABLE OF IR ABSORPTION FREQUENCIES AND NMR PROTON SHIFT VALUES

Class of Compound	Identifying Group	Region of IR Absorption (cm^{-1})	Chemical Shift Region (δ)	
			 (H) (dashed circle)	 (H) (solid circle)
Acids	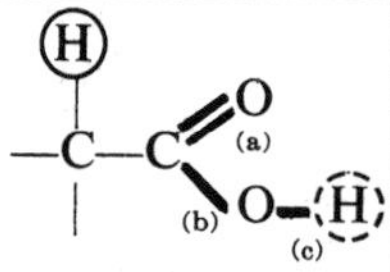 (H)—C—C(=O)(a)—(b)O—(H)(c)	(a) 1680–1725 (b) 1250 (c) 2500–3000	10–12	2–2.6
Alcohols	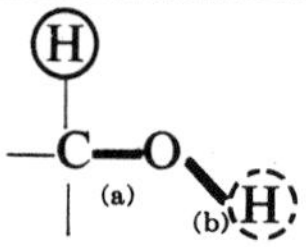 (H)—C—O(a)—(b)(H)	(a) 1000–1200 (b) 3200–3600	1–7	3.4–4
Aldehydes	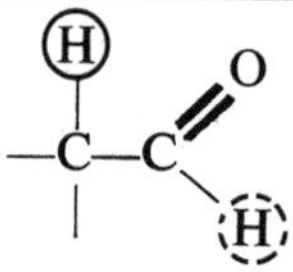 (H)—C—C(=O)—(H)	1690–1760	9–10	2–2.7
Amines	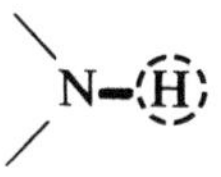 N—(H)	3200–3500 (—NH_2 may appear as a doublet)	1–5	

TABLE 14.2 (*Continued*)

Class of Compound	Identifying Group	Region of IR Absorption (cm^{-1})	Chemical Shift Region (δ)	
			(H) dashed	(H) circled
Aromatic Rings	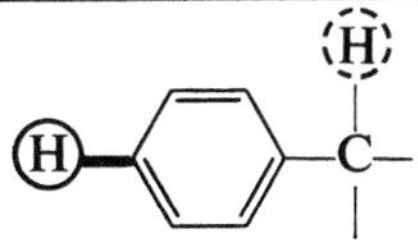 (H)–C₆H₄–C(H)<	675–870 also 3000–3100	2.2–3	7–8
Ethers	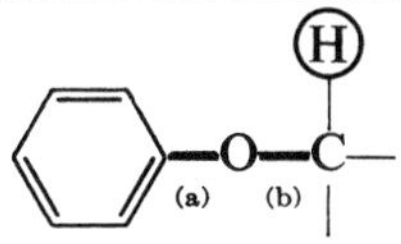 C₆H₅–O–C(H)< (a) (b)	(a) 1200–1275 (b) 1060–1150		3.3–4
Ketones	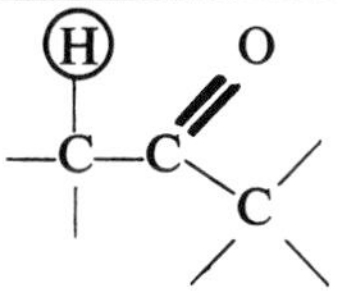 (H)–C–C(=O)–C<	approximately the same as aldehydes		approx. same as alde- hydes
Phenols	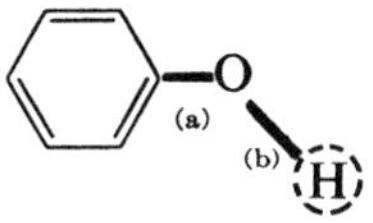 C₆H₅–O–(H) (a) (b)	(a) 1140–1230 (b) 3200–3600	4–12	

TEXT REFERENCE

Report: 14

Chapter Pages

Section Desk NAME

Unknown No.:________

1. Conclusions drawn from ir and nmr spectra studies:

Significant ir Absorption Frequencies (cm^{-1})	Inferences

Significant nmr Chemical Shift Values (δ)	Inferences

2. Physical constants: Melting point________°C
Boiling point ________°C

3. Solubility tests:

H_2O	Ether	10% HCl	10% NaOH	10% Na_2CO_3	85% H_3PO_4

4. Chemical classification tests:

Reagent	Results	Inferences

5. Special tests: (describe)

Identity of unknown compound:________________________

Questions and Exercises

* 1. The carbonyl group shows an ir absorption somewhere within the wavelength range of 5.347–6.134μ. Show how these wavelength values may be expressed in terms of wave numbers within the range of 1870–1630 cm^{-1} respectively.

* 2. In the explanation of ir absorption phenomena, bonded atoms are frequently pictured as small spheres connected by miniature springs, and vibrational frequencies of the bonded atoms likened to that of a simple harmonic oscillator. Refer to your text and, from the Hooke's law equation, tell what mainly determines the characteristic absorption frequency of any functional group.

* 3. An organic compound, $C_9H_{10}O$, may be either an aldehyde or a ketone. What single feature in the nmr spectra of one of these two classes of compounds easily distinguishes it from the other?

* 4. If, on strong oxidation with chromic acid, the compound in Question 3 yielded only benzoic acid, how many isomeric structures could be drawn with this molecular formula? If the nmr spectrum of the compound showed only three singlets, what is its structure and name?

5. If you were issued an unknown liquid, $C_6H_{12}O$, and asked to identify it, outline in a stepwise procedure exactly how you would go about it.

Operation	Result	Inference
(1)		
(2)		
(3)		
etc.		

Experiment 15

Time: 3½ hours

Fats and Oils; Soaps and Detergents

The fats and oils of animal and vegetable origin are **glycerides** or long-chain fatty acid esters of the trihydric alcohol glycerol, $HOCH_2CH(OH)CH_2OH$.

$$\begin{array}{l} CH_2O\text{—}COR \\ | \\ CH\text{—}O\text{—}COR' \\ | \\ CH_2O\text{—}COR'' \end{array}$$

A general formula for a fat or oil

The hydrocarbon segments indicated by R, R′, and R″ in the formula above generally are different. They may vary not only in length but also in degree of unsaturation. If the fatty acid components of a glyceride are long-chain (C_{12}—C_{18}) and saturated, the ester is a solid or a semisolid at room temperature and is classified as a fat. If the long-chain acid residues are unsaturated (i.e., contain one or more double bonds), the glyceride is a liquid at room temperature and is classified as an oil. Unsaturation in the acid components of a fat lowers its melting point. Conversely, saturating the double bonds with hydrogen raises its melting point. The latter process, when applied to oils, is known as hardening or hydrogenation and is carried out as an industrial process for the manufacture of margarines and cooking fats from vegetable oils.

A fat or an oil, when saponified (i.e., hydrolyzed with alkali), produces glycerol and the sodium or potassium salts of a mixture of fatty acids. The latter are called **soaps.**

$$\begin{array}{lcll} CH_2O\text{—}COR & & CH_2OH & RCOO^-\,Na^+ \\ | & & | & \\ CH\text{—}O\text{—}COR' + 3\,NaOH & \rightarrow & CHOH\ + & R'COO^-\,Na^+ \text{ (soaps)} \\ | & & | & \\ CH_2O\text{—}COR'' & & CH_2OH & R''COO^-\,Na^+ \\ \text{A fat or oil} & & \text{Glycerol} & \end{array}$$

The sodium and potassium soaps are soluble in water and are used as cleansing agents. The calcium, magnesium and ferric salts of the same fatty acids are insoluble in water and are not useful as soaps. These insoluble metal salts precipitate as a scum when ordinary soap is used in hard water. **Syndets** (*syn*thetic *det*ergent*s*) do not form insoluble salts with the metallic ions normally present in water.

Acidification of a solution of soap will cause the fatty acid to precipitate.

$$RCOO^{-}Na^{+} + HCl \rightarrow RCOOH + Na^{+}Cl^{-}$$

Procedure

A. SOLUBILITY

Note the odor and appearance of cottonseed oil. Try dissolving a little in water; in alcohol; in methylene chloride. Repeat the test using samples of margarine.

B. UNSATURATION TESTS

Dissolve 0.5 ml of cottonseed oil in 5 ml of methylene chloride in a test tube. Add a solution of 5% bromine in methylene chloride dropwise, counting the number of drops required until the bromine color no longer is discharged instantly. Repeat the test using 0.5 g of Crisco or some other hydrogenated shortening.

C. DRYING OILS

On different parts of a glass cover plate place one drop of each of the following oils: (a) boiled linseed oil, (b) cottonseed oil, (c) Mazola or some other corn oil. Put an identifying mark by each and let them stand in your locker until the next laboratory period. Observe and report the condition of each oil at that time.

D. THE PREPARATION OF SOAP

Support on an iron ring or on a tripod a water bath filled about two-thirds with water, and heat the water to boiling. Dissolve 2.5 g of sodium hydroxide in 5 ml of *distilled* water and 10 ml of ordinary alcohol (95%). Add the alkaline solution to 5 g of Crisco in a 150-ml beaker. Cover the beaker with a watch glass and heat the mixture on the water bath. Stir frequently to prevent spattering and keep the volume of the solution fairly constant by adding small amounts of 50% alcohol. If the mixture foams too much add a small amount of undiluted alcohol. **CAUTION!** *Keep the alcohol away from your burner.* The reaction is complete when the oil or melted fat has dissolved and gives a clear homogeneous solution (about one-half hour). Dilute your soap solution by adding 15 ml of water then pour it into a brine made by dissolving 30 g of sodium chloride in 100 ml of distilled water. Stir the mixture thoroughly and collect the precipitated soap on the Büchner funnel. Wash the soap twice with 10-ml portions of cold distilled water.

Dissolve 2 g of the crude soap in 100 ml of distilled water and set the mixture aside as your test solution. Place the remainder of the soap in an evaporating dish, heat it on the water bath, and stir into the soap just enough water to form a thick solution. Allow the soap solution to cool. Unless the amount of water added was excessive, the soap will solidify into a cake somewhat resembling commercial soap.

Perform the following experiments on samples of your test solution and record your results on the report form.

(a) ALKALINITY

Test the alkalinity of your dilute soap solution with pink litmus paper and compare the result with those of similar tests made on solutions prepared from 0.5 g-samples of Ivory Flakes and Dreft, each dissolved in 50 ml of distilled water. Record your results.

(b) METALLIC SALTS OF FATTY ACIDS

To a 10-ml portion of your soap solution add 1 ml of a dilute (0.1%) calcium chloride solution. Shake vigorously and observe. Repeat the test with dilute magnesium chloride and ferric chloride solutions. Perform the same tests on the samples of Ivory and Dreft solutions prepared in (a). Record your results.

(c) PRECIPITATION OF SOAP

To 50 ml of your soap solution add dilute hydrochloric acid dropwise until the solution is acid to Congo Red paper. Cool the mixture in ice, collect the precipitated acid on your suction funnel, and wash it with 20 ml of cold water. On your report form write an equation to show the reaction which took place. Assign a name to your product and show it to your laboratory instructor when handing in your written report.

Test the solubility of a small sample (0.25 g) of your product in 2 ml of methylene chloride. Also test the solubility of a small sample of stearic acid in the same solvent.

(d) EMULSIFYING ACTION OF SOAP

Shake 4 drops of mineral oil with 10 ml of soap solution. Repeat the experiment using 10 ml of water. Reexamine both mixtures after standing for 5 minutes. Record your results.

TEXT REFERENCE

Report: 15

Chapter Pages

Section Desk NAME

FATS AND OILS; SOAPS AND DETERGENTS

SOLUBILITY OF FATS AND OILS

Solvent used	Observation
Water	
Alcohol	
CH_2Cl_2	

UNSATURATION TESTS

Number of drops of 5% bromine in methylene chloride required to saturate your sample of cottonseed oil. ________

Number of drops of 5% bromine in methylene chloride required to saturate a sample of Crisco. ________

Draw the structure of the principal glyceride found in cottonseed oil (consult text) and write the equation for the reaction which took place with bromine.

DRYING OILS

Describe the appearance of your oil samples after standing one week.

Boiled linseed oil. ______

Cottonseed oil. ______

Mazola or other corn oil. ______

SOAPS AND DETERGENTS

Tests performed	Your soap (brand X)	Detergents	
		Ivory	Dreft
(a) Alkalinity (litmus)			
(b) $CaCl_2$ solution			
$MgCl_2$ solution			
$FeCl_3$ solution			

(c) Equation for reaction with HCl

Solubility of product in CH_2Cl_2. ______

Solubility of stearic acid in CH_2Cl_2. ______

(d) Emulsifying action of soap on oil.

Oil and soap mixture. (Result) ______

______

Oil and water. (Result) ______

Questions and Exercises

1. Draw formulas and give names for glycerides which probably are present in lard; in Mazola oil.
2. Could cottonseed oil be used in paints as a drying oil? Explain.
3. Assume the average molecular weight of a fat is 890. What fatty acid components probably predominate?
4. Why was the dilute soap solution poured in a solution of sodium chloride in water.
5. How does soap function as an emulsifying agent for oil in water? Draw a picture of an oil droplet and several soap molecules to illustrate your answer.
6. Suppose that you wanted to emulsify a water-insoluble compound in water. Would soap be a reasonable choice for the emulsifying agent if the water were slightly acidic? Why? If it would not be a good choice, can you draw the structure of an organic molecule that might be more suitable?
7. If castor oil (Experiment 29) were treated with sulfuric acid and then neutralized with sodium hydroxide, what would probably be the result? Would this structure be water-soluble?
8. Saponification (conversion into soap) is a term also used to describe the alkaline hydrolysis of any ester. Why is alkaline hydrolysis of an ester to be preferred over acid hydrolysis?

Experiment 16

Time: 3½ hours

Carbohydrates

For convenience the carbohydrates may be divided into three classes:

(1) **Monosaccharides** (polyhydroxy aldehydes or ketones), which do not yield smaller units when hydrolyzed.

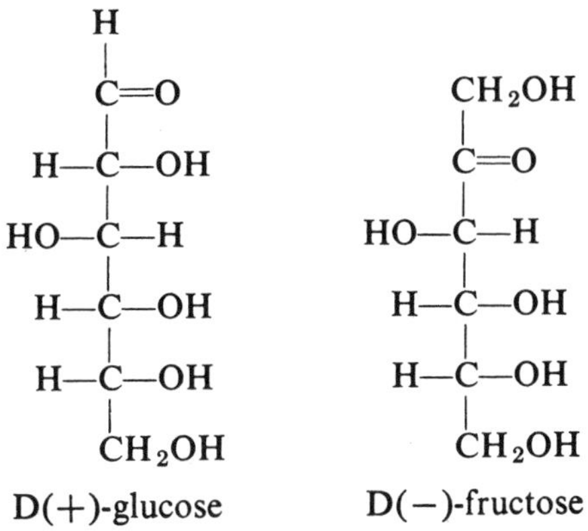

```
       H
       |
       C=O                 CH2OH
       |                    |
   H—C—OH                 C=O
       |                    |
  HO—C—H              HO—C—H
       |                    |
   H—C—OH              H—C—OH
       |                    |
   H—C—OH              H—C—OH
       |                    |
      CH2OH                CH2OH
```

D(+)-glucose   D(−)-fructose

(2) **Disaccharides,** which yield, when hydrolyzed, two monosaccharide units.

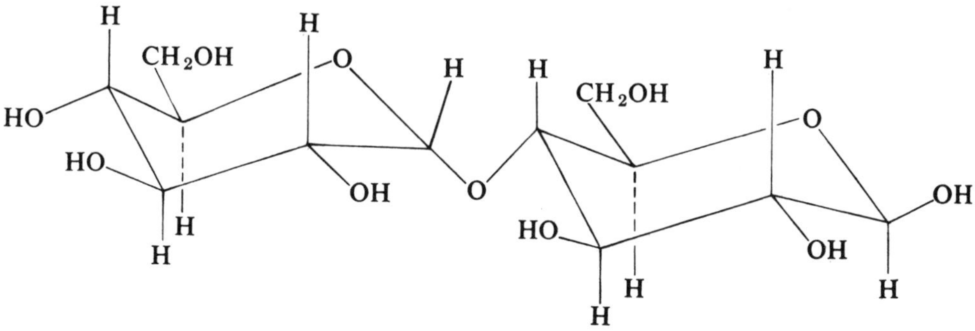

Maltose, a reducing disaccharide yields two glucose units.

(3) **Polysaccharides,** which, when hydrolyzed, yield many molecules of monosaccharides.

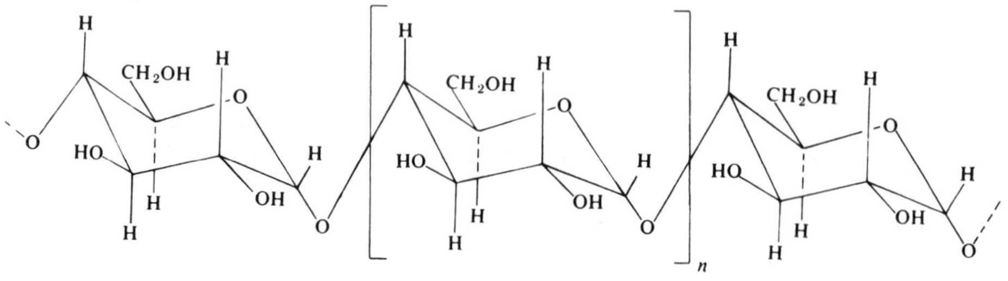

A small segment of the starch molecule (n = 6,000–30,000)

The following experiments illustrate some of the typical reactions of each class.

Procedure

A. PREPARATION OF SUGAR SOLUTIONS

Prepare 4% solutions of the following sugars by dissolving 1 g of each in 25 ml of distilled water: glucose, fructose, sucrose, maltose, and lactose. Use these test solutions in procedures (B–D) inclusive. Record all your observations on the report form and explain each test. Use equations whenever possible.

B. THE MOLISCH TEST

All substances having a carbohydrate grouping react with Molisch's reagent to give a purple color. The reaction is quite complex, but the test is a reliable indication of the presence of carbohydrates. Test one of the sugar solutions prepared in procedure (A) as follows:

To 2 ml of the sugar solution in a test tube add 2 drops of Molisch's reagent (15% solution of α-naphthol in ethanol) and mix by shaking. Pour this solution slowly down the inside of a second test tube containing 2 ml of concentrated sulfuric acid. Hold the second test tube at an angle of approximately 30° while pouring so that two separate layers are formed without mixing. Observe any color change at the junction of the two liquids. Repeat the test using a suspension made by mixing a small microspatula measure of flour in 2 ml of water.

C. BENEDICT'S OR FEHLING'S TEST FOR REDUCING SUGARS

All monosaccharides reduce Benedict's or Fehling's solutions whether they are aldehyde sugars (aldoses) or ketone sugars (ketoses). Disaccharides, on the other hand, are subdivided into two classes on the basis of their behavior toward these reagents as: (a) **reducing disaccharides,** that reduce Benedict's or Fehling's solutions, and (b) **nonreducing disaccharides,** that do not react with these solutions.

Test each of the sugar solutions prepared in procedure (A) separately in the following manner:

Place 4 ml of Benedict's solution or 4 ml of freshly prepared Fehling's solution[1] in a test tube and heat the solution to gentle boiling. To the boiling solution add the sugar solution 2–3 drops at a time, heating the mixed solutions for at least one minute after each addition. Observe any color changes and continue adding the sugar solution and heating until the blue color just disappears, but do not add more than 5 ml of any sugar solution. Note the appearance and color of any precipitate.

D. OSAZONE FORMATION

Monosaccharides and reducing disaccharides react with phenylhydrazine to yield **osazones.** These are condensation products which contain *two* phenylhydrazone groups rather than one as would be expected from the reaction of a simple aldehyde or ketone. The osazones are crystalline derivatives useful in the identification of the sugars. Each osazone possesses a definite melting point and crystalline form, and the length of time required to form the osazone is a characteristic of the sugar being tested.

Support over a wire gauze a 400-ml beaker half-filled with water, and heat the water to boiling. Label or mark 5 test tubes and to each add 4 ml of phenylhydrazine reagent [Experiment 12(A)] and 2 drops of saturated sodium bisulfite solution. To each test tube add 5 ml of the various sugar solutions prepared in procedure (A) and label each test tube with the sugar it contains. Mix the solutions throughly and immerse all tubes at the same time in the beaker of boiling water. Record the time of immersion and the time at which each osazone begins to precipitate (Note 1). Shake the tubes from time to time to prevent formation of a supersaturated solution of the osazone, and continue to heat the tubes for twenty minutes. Allow the tubes to cool slowly by removing them from the water bath and placing them in the test tube rack. Filter any crystals formed on the Hirsch filter and examine them. Allow the osazone of one sugar to dry and determine its melting point. If a low power

[1] See appendix.

microscope is available, transfer a few osazone crystals to a microscope slide and examine them. Are they all alike? Sketch and name the osazone.

E. HYDROLYSIS OF DISACCHARIDES AND POLYSACCHARIDES

(a) INVERSION (HYDROLYSIS) OF SUCROSE

Place the remainder of the sucrose solution prepared in procedure (A) in an Erlenmeyer flask, add 1 ml of dilute hydrochloric acid, and heat the mixture on the water bath for thirty minutes. Cool the solution and carefully neutralize it with 10% sodium hydroxide (use litmus). Test the neutralized solution with Benedict's or Fehling's solution as in procedure (C).

(b) STARCH (ACID HYDROLYSIS)

Place a beaker containing 120 ml of distilled water on a wire gauze and heat the water to boiling. Mix 1 g of starch thoroughly with 10 ml of cold water, stirring and crushing until the suspension is free of lumps. Pour the suspension slowly with stirring into the boiling water and boil the mixture for 1–2 minutes after addition is complete. Test a portion of the starch solution with Benedict's reagent. Also test a small sample of the starch solution with a drop of iodine-potassium iodide solution.[1]

To 50 ml of the starch solution in a 125-ml Erlenmeyer flask add 5 drops of concentrated hydrochloric acid and heat the mixture for 30–40 minutes on the water bath. Cool the solution and carefully neutralize it with 10% sodium hydroxide. Test a portion of the cooled solution again with Benedict's solution and also with the iodine-potassium iodide solution. Result?

(c) BY ENZYMES

To 50 ml of the starch solution in a 125-ml Erlenmeyer flask add about 2 ml of your own saliva. Mix and place in a warm (40°) water bath for 30 minutes. Again test portions of the solution with Benedict's reagent and with iodine-potassium iodide solution. Result?

F. ACETYLATION OF CELLULOSE

Cellulose (a polysaccharide) is also a polyhydric alcohol and may be esterified with acetic anhydride to give **cellulose acetate,** a compound in which two to three of the hydroxyl groups in each monosaccharide unit have been acetylated. Cellulose acetate is used commercially in the manufacture of camera film, rayon, adhesives, etc.

In a 50-ml Erlenmeyer flask add 20 ml of glacial acetic acid, 7 ml of acetic anhydride, and 2 drops of concentrated sulfuric acid. Press 0.5 g of cotton beneath the surface of the liquid mixture with a stirring rod, making sure that the cotton is completely immersed. Allow the mixture to stand until the next laboratory period or heat it at 70–75° in a water bath for 30–45 minutes with occasional stirring.

[1]See appendix.

Pour the acetylated cotton and solution with stirring into 300 ml of cold water. Filter the gelatinous precipitate with gentle suction on the Büchner funnel. Wash the precipitate with 100 ml of water, press between two sheets of filter paper to remove as much water as possible, and air dry. Dissolve about one fourth of the thoroughly dry product in a mixture of 15 ml of acetone and 0.5 ml of ethanol. Add 1–2 drops of dibutyl phthalate to the solution and pour the solution onto a watch glass. Allow the solvent to evaporate. Pour water onto the film and allow to stand for a few minutes. Carefully separate the film from the glass and examine. Try burning a small sample. Report your observations and attach a small sample of your product to your report sheet.

G. NITRATION OF CELLULOSE

In a 100-ml beaker mix, *cautiously*, 10 ml of concentrated sulfuric acid, and 10 ml of concentrated nitric acid. Heat the solution on the hot water bath to 50° and then immerse a wad of absorbent cotton weighing about 0.5 g in the solution. Stir occasionally and allow the cotton to remain in this nitrating mixture no longer than 4 minutes. After removing the cotton, immediately immerse it in a large beaker of cold water. Change the water several times until a final rinsing no longer produces a wash water acid to litmus. Squeeze the cotton as dry as possible and spread out on filter paper to dry. Cotton partially nitrated in this manner produces **pyroxylin.** Hold a piece of ***dry*** pyroxylin by means of your tongs in the flame of a burner. Result?

Dissolve a portion of the pyroxylin in 20 ml of a 50-50 mixture of alcohol and ether. Puddle and stir until completely dissolved. Decant onto a watch glass and allow the solvent to evaporate (Note 2). Describe the appearance of the residual film. Lift the edge of the film and allow water to run beneath it to facilitate easy removal. Attach a sample to your report form.

Note 1. The osazones of lactose and maltose frequently form only upon cooling the reaction mixture. The partial hydrolysis of sucrose through long periods of heating results in the formation of sufficient glucose and fructose to give a positive osazone reaction. A positive reaction therefore is often erroneously reported for sucrose.

Note 2. Enough nitrated cotton should be dissolved so the mixture will pour onto the watch glass as a syrupy liquid.

TEXT REFERENCE

Report: 16

Chapter Pages

Section Desk NAME

CARBOHYDRATES

RESULTS OF MOLISCH TEST

Test with:

Glucose	Fructose	Sucrose	Maltose	Lactose	Starch

RESULTS OF THE BENEDICT'S OR FEHLING'S TEST

Test with:

Glucose	Fructose	Sucrose	Maltose	Lactose

Write an equation for the reaction which takes place when glucose is heated with Fehling's solution.

RESULTS OF THE OSAZONE TEST

Test with:

Glucose	Fructose	Sucrose	Maltose	Lactose

Give time required for each reaction

If osazone crystals were viewed through a microscope, make a sketch of the crystals. Name the osazone.

Color? ____________________

mp ____________________
(experimental)

mp ____________________
(literature)

Sketch of crystal

RESULTS OF HYDROLYSIS

(a) Inversion of sucrose. Describe your results with Benedict's or Fehling's solutions.

____

____

____.

(b) Acid hydrolysis of starch.

Test with Benedict's reagent	
Before hydrolysis__________	Pos. or Neg.__________
After hydrolysis __________	Pos. or Neg.__________

Test with iodine-potassium iodide	
Before hydrolysis__________	Pos. or Neg.__________
After hydrolysis __________	Pos. or Neg.__________

(c) Hydrolysis catalyzed by enzymes.

Test with iodine-potassium iodide   Positive or Negative ________

Test with Fehling's Solution   Positive or Negative ________

CELLULOSE ACETATE AND CELLULOSE NITRATE

Describe the appearance of these films ________________________

____

Describe the difference in burning properties of cellulose nitrate and cellulose acetate. ________________________

(Attach samples here)

Questions and Exercises

1. Why are the osazones of glucose and fructose identical?
2. Which other pairs of aldohexoses besides glucose and mannose give identical osazones?
3. Draw the structure for the disaccharide sucrose and point out why it is not a reducing sugar.
4. Why is the hydrolysis of sucrose referred to as "inversion"?
5. Write a mechanism for the acid hydrolysis of sucrose.
6. To what general class of compounds does cellulose acetate belong? Cellulose nitrate? Ethyl cellulose?
7. What is the maximum number of hydroxyl groups per glucose unit of cellulose that can be nitrated? If all these are nitrated, what commercial product is formed?
8. One of the earliest synthetic plastics, called *Celluloid*, was manufactured from cellulose nitrate and camphor. Celluloid is now seldom seen. What obvious disadvantage keeps celluloid from being used in place of some of our modern plastics?

Experiment 17

Time: 2 × 3 hours

Proteins

Proteins are very complex, high molecular weight (10,000 to 10,000,000) structures that yield α-amino acids on hydrolysis.

$$\begin{array}{cccccccccccccccc}
 & H & & R & & O & \vdots & H & & R' & & O & \vdots & H & & R'' & & O \\
 & | & & | & & \| & \vdots & | & & | & & \| & \vdots & | & & | & & \| \\
— & N & — & C & — & C & — & N & — & C & — & C & — & N & — & C & — & C & — & + \; 3\,H_2O \rightarrow \\
 & & & | & & & \vdots & & & | & & & \vdots & & & | \\
 & & & H & & & \vdots & & & H & & & \vdots & & & H
\end{array}$$

A portion of a protein molecule

$$\begin{array}{c}
 \quad H \;\; O \\
 \quad | \;\;\; \| \\
R—C—C—OH \\
 \quad | \\
 \quad NH_2
\end{array}
+
\begin{array}{c}
 \quad H \;\; O \\
 \quad | \;\;\; \| \\
R'—C—C—OH \\
 \quad | \\
 \quad NH_2
\end{array}
+
\begin{array}{c}
 \quad H \;\; O \\
 \quad | \;\;\; \| \\
R''—C—C—OH \\
 \quad | \\
 \quad NH_2
\end{array}$$

α-Amino acids

The hydrolysis of a protein may be catalyzed by the action of enzymes, acids, or bases and may yield mixtures of as many as twenty or more α-amino acids.

Procedure

A. HYDROLYSIS OF A PROTEIN

Assemble a small-scale reflux apparatus as illustrated in Figure 2.3. Introduce into the boiling flask 20 ml of 20% hydrochloric acid (concentrated hydrochloric acid is 36%), 0.5 g of casein, and a piece of porous plate or a "Boileezer" or two. Heat the mixture under reflux for 30–45 minutes using a very low flame. Control the heating very carefully until the casein is dissolved. While the mixture is refluxing, proceed with procedures (B), (C), (D), and (E). After hydrolysis is complete (as determined by tests (a) and (b) below), cool the mixture, stopper, and save for procedure (H). If hydrolysis is not complete after this reflux period (as indicated by test (a)), add 3–5 ml additional 20% hydrochloric acid and heat for an additional 15 minutes or until test (a) is negative.

(**a**) Carefully neutralize (using litmus) 2–3 ml of the hydrolysate with 10% sodium hydroxide solution. Do the amino acids precipitate? Make the solution definitely alkaline by adding an additional 1 ml of dilute sodium hydroxide. Add 1–2 drops of 2% copper sulfate solution. Do you obtain a positive biuret test? [See procedure (E)].

(**b**) Amino acids, like other primary aliphatic amines, react with nitrous acid to liberate gaseous nitrogen.

$$\underset{\text{An }\alpha\text{-amino acid}}{R{-}\overset{\overset{\displaystyle H}{|}}{\underset{\underset{\displaystyle NH_2}{|}}{C}}{-}\overset{\overset{\displaystyle O}{\|}}{C}{-}OH} + HNO_2 \rightarrow \underset{\text{An }\alpha\text{-hydroxy acid}}{R{-}\overset{\overset{\displaystyle H}{|}}{\underset{\underset{\displaystyle OH}{|}}{C}}{-}\overset{\overset{\displaystyle O}{\|}}{C}{-}OH} + H_2O + N_2$$

To 5 ml of your acidic protein hydrolysate slowly add an equivalent volume of 5% sodium nitrite. Repeat the test using a solution of 0.1 g of glycine in 5 ml of 10% hydrochloric acid. Record your observations.

B. TEST FOR NITROGEN

Mix thoroughly, in a small evaporating dish, 0.5 g of casein or dried egg albumin with 1–2 g of finely powdered soda lime (sodium hydroxide and calcium oxide). Transfer the mixture to a dry test tube and heat over a low flame. Note cautiously the odor of the gas which evolves and test it with a piece of moist red litmus held over the mouth of the test tube. What was the product formed?

C. TEST FOR SULFUR

To about 0.5 g of dried egg albumin in a 50-ml Erlenmeyer flask add 10–15 ml of 10% sodium hydroxide and very gently boil for about 15 minutes. The mixture will have a tendency to froth easily. Cool the solution and make it acidic with hydrochloric acid. Again bring to a boil after having placed a piece of moist lead acetate paper over the mouth of the flask. Observe the change in the acetate paper. What material is formed? What amino acids give a positive test?

D. XANTHOPROTEIC TEST

Proteins which have α-amino acid units containing aromatic nuclei may be nitrated to give yellow compounds, which deepen in color when treated with alkali. You have unwittingly performed this test if you ever have spilled nitric acid on your fingers.

To 3 ml of egg-white solution[1] in a test tube add 10–15 drops of concentrated nitric acid and warm gently. Observe and record any color change.

[1]An egg-white solution is prepared by shaking vigorously the white of an egg with 100 ml of water in a stoppered 250-ml Erlenmeyer flask.

Cool the solution and neutralize it with 10% sodium hydroxide. Note and record any change in color.

E. THE BIURET TEST

The biuret reaction is a general test for proteins and for compounds which have in their structures certain multiple amide linkages. Perhaps the simplest compound to give a positive test is biuret itself, $H_2N{-}\overset{\overset{\large O}{\|}}{C}{-}\overset{\overset{\large H}{|}}{N}{-}\overset{\overset{\large O}{\|}}{C}{-}NH_2$. A positive biuret test is indicated by the production of a violet-pink color when alkaline solutions of compounds containing the above structural units are treated with a dilute solution of copper sulfate.

In each of two dry test tubes place 1 g of urea. Heat one test tube gently until the urea melts. Note cautiously the odor of the gas which evolves and test it with a piece of moist red litmus held over the mouth of the test tube. Continue heating gently until the material shows a tendency to solidify. Dissolve the white residue in 3–4 ml of warm water and filter. To the filtrate add an equal volume of 10% sodium hydroxide and mix. To this alkaline solution add 1–2 drops of 2% copper sulfate. Shake the mixture and observe the color. Record your results and write the equation for the formation of biuret.

Without heating the other sample of urea, add 3–4 ml of water, make alkaline, and add copper sulfate as before. Result?

F. AMPHOTERIC NATURE OF PROTEINS

Into each of two test tubes place 5 ml of egg-white solution. To one test tube add 2 ml of 10% sodium hydroxide and to the other add 2 ml of 10% hydrochloric acid. Warm the contents of both test tubes on the water bath for about 30 minutes. Cool, then neutralize the acid solution (use litmus) by adding dropwise 10% sodium hydroxide, and neutralize the alkaline solution by adding dropwise 10% hydrochloric acid (use litmus). In each case what occurs at the neutral point? Remove the precipitate from each test tube by filtration and test in the following manner: (1) transfer a small portion of the precipitate to a test tube, add 2–3 ml of 10% hydrochloric acid, and shake well. Result? (2) Place a small portion of the precipitate into a second test tube and add 2–3 ml of 10% sodium hydroxide solution. Shake well and again observe your result. (3) In a third test tube shake a small portion of the precipitate with 2–3 ml of distilled water. Result?

G. PRECIPITATION WITH SALTS OF HEAVY METALS

Into each of four test tubes place 2–3 ml of the egg-white solution prepared in procedure (D). To one test tube add slowly 10% mercuric chloride solution. (**CAUTION!** *Mercuric chloride is very poisonous.*) Observe the result. Repeat the test by adding slowly 10% solutions of ferric chloride, lead acetate, and copper sulfate. Record your results.

H. THE SEPARATION OF α-AMINO ACIDS BY PAPER CHROMATOGRAPHY

The technique used for the separation of α-amino acids is similar to that described in Part II, Experiment 6, but somewhat more refined.

A paper of the same type and size as previously described is used, but special care must be taken to handle the paper only by its upper end (Note 1). Place the paper on a clean sheet of notebook paper, draw a *pencil* line as before, but make six marks at 1.5 cm intervals. Identify these marks as A, P, T, M, and H. Using a separate micropipette for each sample, place at M a small (1.0 mm) spot of a mixture of known amino acids made by mixing equal volumes of 0.1 molar solutions of aspartic acid, phenylalanine, and tyrosine. At A, P, and T place spots of aspartic acid, phenylalanine and tyrosine solutions respectively. At H place a small spot of the hydrolysate saved from procedure (A). Allow the spots to dry and then spot them a second time to make certain sufficient material will be present for an easily observed resolution. When the spots are dry, position the paper in a 500-ml Erlenmeyer flask as before and secure it (Note 2). Through a long stemmed funnel, transfer to the bottom of the 500-ml Erlenmeyer flask 15 ml of a freshly prepared 80% phenol solution. (**CAUTION!** *Phenol has a caustic effect upon the skin. Should you spill phenol on your fingers or hands, immediately wash with copious quantities of soap and water.*) Close the mouth of the flask with aluminum foil and allow the solvent to rise up the paper for approximately $1\frac{1}{2}$ hours. At the end of this time remove the paper and mark the upper limit of the solvent front. Wash off the excess phenol by rinsing the paper thoroughly on both sides with acetone from a squeeze bottle. After the acetone has evaporated, spray the paper lightly with a 2% solution of ninhydrin in 95% ethyl alcohol. The paper must be thoroughly sprayed with ninhydrin but should not be dripping wet. Hang the paper up to dry. Colored spots will begin to appear within 5–15 minutes. Circle each colored spot and measure the distance from its center to the starting line. Calculate and record all R*f* values for each amino acid on your report form. Compare your R*f* values with those given in Table 17.1 (Note 3).

Note 1. Fingerprints will show up as colored spots when the chromatogram is developed if the paper is carelessly handled. Plastic gloves of the throw-away type worn during the preparation of the chromatogram and the spraying operation will prevent both the paper and hands from unnecessary coloration.

Note 2. It is very important that the spots be dry before solvent moves over them. If not dry, spots become diffuse.

Note 3. Easy to use aerosol sprays are commercially available for α-amino acid chromatography (see Appendix), but a "Windex" glass cleaner bottle or a small insecticide sprayer may be used.

TABLE 17.1

APPROXIMATE PERCENTAGE COMPOSITION AND
R*f* VALUES OF AMINO ACIDS IN CASEIN

Amino acid	Formula	%	R*f* value
Cystine	$S—CH_2CH(NH_2)COOH$ | $S—CH_2CH(NH_2)COOH$	0.35	0.16
Aspartic acid	$HOOC—CH_2CH(NH_2)COOH$	7.1	.32
Glutamic acid	$HOOC—CH_2CH_2CH(NH_2)COOH$	23.3	.40
Glycine	$H_2N—CH_2COOH$	2.7	.42
Serine	$HOCH_2CH(NH_2)COOH$	7.7	.43
Threonine	$CH_3CH(OH)CH(NH_2)COOH$	4.9	.51
Alanine	$CH_3CH(NH_2)COOH$	3.0	.59
Tyrosine	HO—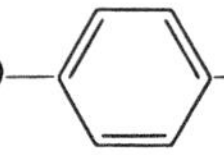$—CH_2CH(NH_2)COOH$	6.3	.62
Lysine	$H_2N(CH_2)_4CH(NH_2)COOH$	8.2	.71
Valine	$(CH_3)_2CHCH(NH_2)COOH$	7.2	.75
Arginine	NH ‖ $H_2N—C—NH(CH_2)_3CH(NH_2)COOH$	4.1	.76
Methionine	$CH_3S(CH_2)_2CH(NH_2)COOH$	3.4	.77
Leucine	$(CH_3)_2CHCH_2CH(NH_2)COOH$	9.2	.79
Phenylalanine	$—CH_2CH(NH_2)COOH$	5.0	.82
Proline	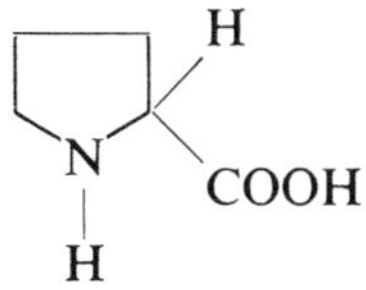	11.3	.85

TEXT REFERENCE

Report: 17

Chapter Pages

Section Desk NAME

PROTEINS

HYDROLYSIS OF A PROTEIN

(a) Result of the biuret test: ______

(b) Reaction with nitrous acid: ______

(c) Reaction of glycine with nitrous acid: ______

TEST FOR NITROGEN

Nature of the product formed ______

Complete the following equation.

$$R{-}\overset{\overset{\displaystyle O}{\|}}{C}{-}NH_2 + NaOH \rightarrow$$

TEST FOR SULFUR

What gas evolved when the acidified albumin hydrolysate was heated?

______

Write the equation for its reaction with lead acetate, $\left(CH_3{-}\overset{\overset{\displaystyle O}{\|}}{C}{-}O{-}\right)_2 Pb$.

XANTHOPROTEIC TEST

Describe the color change which occurred when a protein sample was treated with nitric acid. ______

Describe the color change after addition of sodium hydroxide. ______

______

______

THE BIURET TEST

Describe your results when a heated sample of urea was tested for biuret.

____

What was the result when the unheated sample of urea was tested?

____

Write the chemical equation for the preparation of biuret from urea.

AMPHOTERIC NATURE OF PROTEINS

Sample	Result when neutralized
Egg white + 10% HCl	
Egg white + 10% NaOH	

What was the result when precipitated egg white was treated with acid?

____

When treated with alkali? _______________

ACTION OF HEAVY METAL SALTS ON PROTEINS

Metal salt	Action on egg white
$FeCl_3$	
$(CH_3—\overset{\overset{\displaystyle O}{\|}}{C}—O)_2Pb$	
$HgCl_2$	
$CuSO_4$	

THE SEPARATION AND IDENTIFICATION OF α-AMINO ACIDS (R*f* VALUES)

Amino acid	Experi-mental	Literature
Aspartic acid	________	________
Phenylalanine	________	________
Tyrosine	________	________

(Attach chromatogram here.)

Questions and Exercises

1. Why did the sample of casein hydrolysate give a continuous streak rather than separate spots?
2. What explanation can you offer for the fact that on the casein hydrolysate chromatogram denser areas occur at R*f* values of 0.30–0.45 and again at 0.80–0.85? (Hint: Consult Table 17.1.)
3. Write the structures of three sulfur-containing α-amino acids. Name each.
4. Write the structures of three α-amino acids which would give a color with nitric acid. Name each.
5. Show four different structural arrangements which, if present in a compound, would give a positive biuret test.
6. Why is egg white frequently suggested as an antidote for certain types of heavy metal poisoning?
7. What is a peptide? Write the structure for glycylglycylalanine. To what classification does it belong?
8. Assign a name to the following structure:

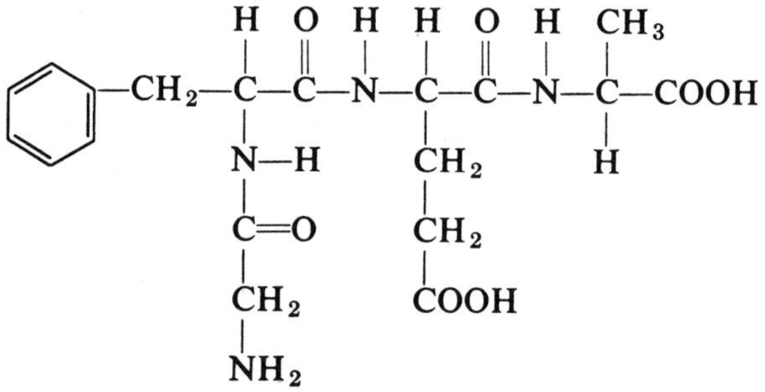

```
                  H  O  H  H  O  H  CH3
                  |  ‖  |  |  ‖  |  |
(C6H5)—CH2—C——C——N——C——C——N——C—COOH
                  |           |           |
                 N—H         CH2          H
                  |           |
                 C=O         CH2
                  |           |
                 CH2         COOH
                  |
                 NH2
```

9. What protein readily available in every grocery store could have served for the preparation of a hydrolysate in Part A?
10. Which of the α-amino acids listed in Table 17.1 would have made up a considerable percentage of the total had we used a hydrolysate prepared from wool?

Experiment 18

Time: 3 + 2 hours

A Biosynthesis

Part I The Preparation of Ethanol

A biochemical synthesis of ethanol, practiced since antiquity and of great commercial importance today, is that achieved through the fermentation of starches and sugars. Fermentation is a very complex, multistep biochemical process in which relatively large structures such as starches are broken down into simple sugars through the catalytic action of enzymes—specific enzymes being required for each step of the degradation. Every brewery, winery, and bakery is dependent upon the chemical changes brought about by fermentation. Yeast is used in the baking process to provide the enzymes (Greek *en*, in; *zyme*, leaven) that initiate fermentation for the primary purpose of providing the leavening action created by the evolution of carbon dioxide. Volatile organic compounds, of course, are removed during the baking process. When a yeast fermentation is carried out for nonbaking purposes, the result is a final potpourri that includes, in addition to ethanol, some aldehydes, ketones, acids, vitamins, minerals, and numerous other components.

It is beyond the scope of this manual to outline here the numerous interacting biochemical reactions involved in the enzymatic conversion of carbohydrates into the various organic substances cited above. For our purpose we shall simply indicate the fermentation process as a chemical reaction which stoichiometrically, at least, takes place according to the following equation:

$$C_6H_{12}O_6 \xrightarrow{\text{Zymase}} 2\,C_2H_5OH + 2\,CO_2$$

As starting material we shall use a simple hexose such as glucose (grape sugar) or fructose (fruit sugar). The sugar in any naturally sweet substance may be fermented, but for our preparation we will begin with freshly prepared apple cider. No yeast is needed because the required enzymes are already present.

According to the preceding equation the fermentation reaction that leads to ethanol requires no oxygen but does produce carbon dioxide. Therefore, the reaction must be carried out under anaerobic conditions in a partially closed system with provision made for allowing carbon dioxide to escape but no air to enter. Without this precaution, acetic acid (vinegar) would be our end product as we shall see in Part II of the experiment. Indeed, if oxygen were freely available only carbon dioxide and water would be produced.

Procedure

Fill a 250-ml Erlenmeyer flask or a pop bottle to within 2 inches of the top with freshly prepared apple cider (Note 1). Stopper the flask with a one-hole rubber stopper through which a section of bent glass tubing may be inserted and led into a CO_2 trap as illustrated in Figure 18.1. Set the cider in a warm place (25°C) and allow fermentation to proceed for 2–3 weeks or until there is no further evolution of carbon dioxide. Filter the product to remove some of the sediment that accumulates during the fermentation, then carry out the following experiments.

FIGURE 18.1 Assembly for carrying out a fermentation on a small scale. (a) Fermenter: (b) carbon dioxide trap.

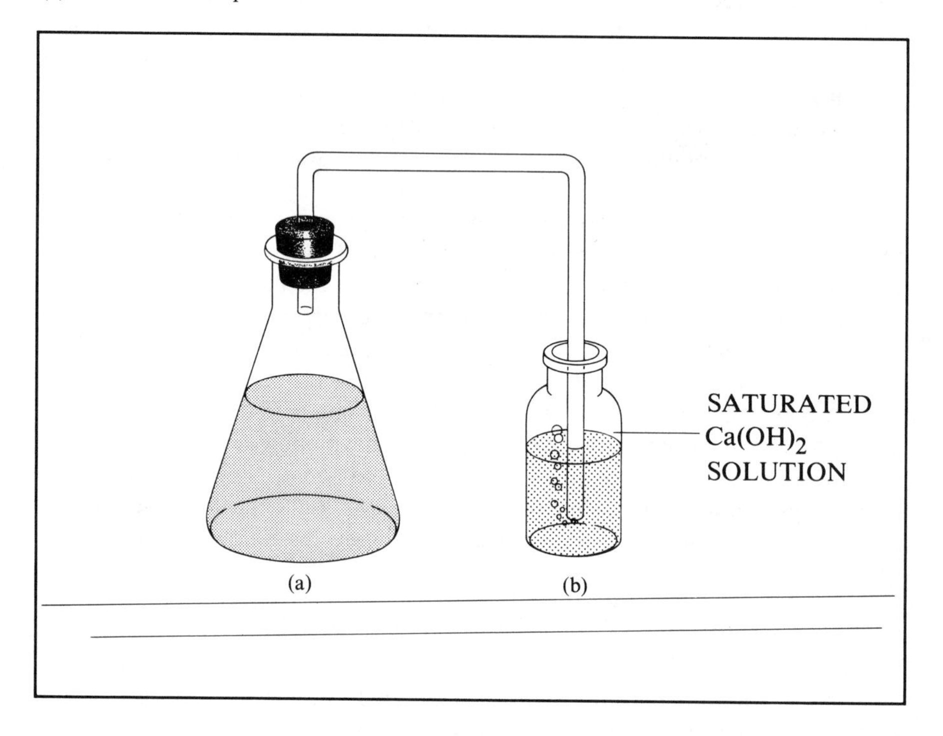

A. GLC ANALYSIS OF ETHANOL

Prepare a fractional distillation assembly (see Figure 4.1) using a 500-ml round-bottomed flask and place in it the entire 250 ml of hard cider filtrate. Add a "Boileezer" and, using a Bunsen burner as your heat source, collect 35–50 ml of distillate in a 100-ml receiving flask. Discard the residual liquid in the boiling flask and replace the latter with the 100-ml flask containing the 35–50 ml of distillate just collected. Using the same fractionating column, redistill and collect only distillate that can be brought over under a temperature of 80°C (*ca.* 10–15 ml).

Using the gas chromatograph, determine the concentration of ethanol in the fraction that you have collected. The general procedure described in Experiment 6-IV may be followed. A column packed with Porapak-S and maintained at a temperature of 150°C and a flow rate of 60 ml/second will be satisfactory. Use 1-μl samples and adjust the recorder attenuation to give peaks of maximum height. Also analyze and record on the same chromatogram a sample of absolute (200 proof) ethanol from the supply room. Compare the chromatograms and, using the supply-room sample as a standard, calculate the percentage composition of the fraction you have collected. Record your calculations on the report form and attach chromatograms. Your chromatogram should resemble that shown in Figure 18.2.

FIGURE 18.2 GLC chromatograms of fermented cider distillate and 200 proof ethanol. Point of injection of samples indicated by (↑).

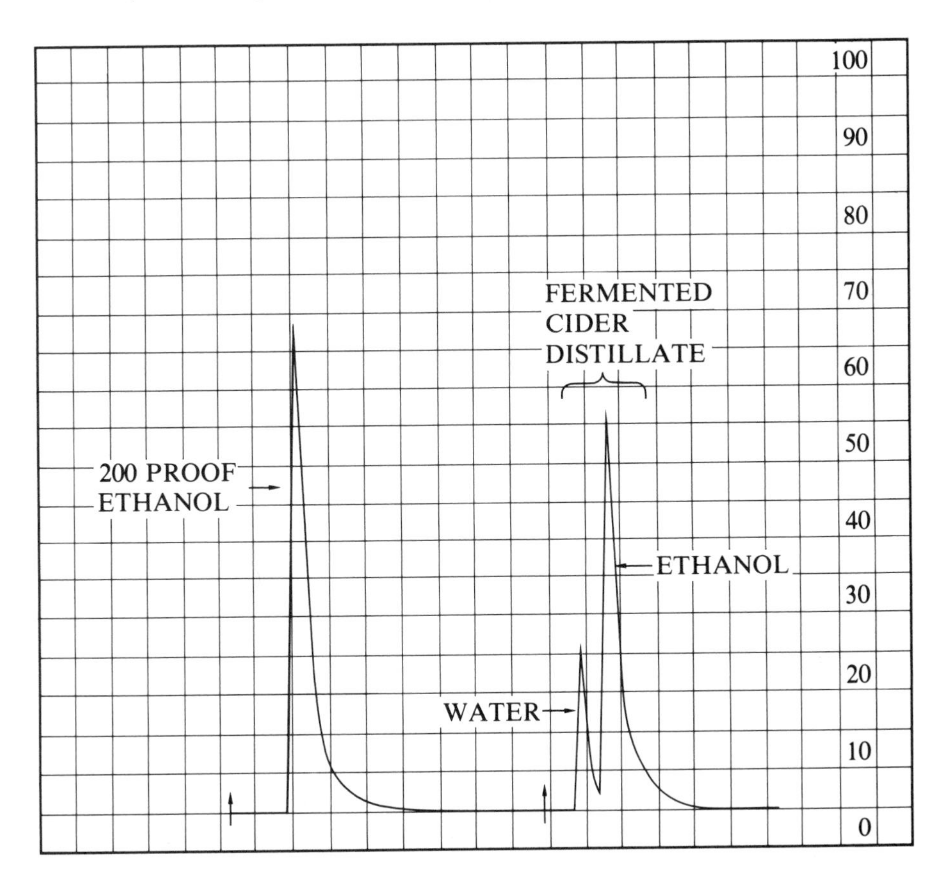

B. OXIDATION TEST

Follow the procedure outlined under Experiment 10 (Part I-D) and carry out the oxidation test on a sample of cider alcohol. Result?

C. IODOFORM TEST

Follow the procedure outlined under Experiment 10 (Part I-E) and carry out the iodoform test on a sample of cider alcohol. Result?

D. ESTERIFICATION

In a test tube place 2 ml of glacial acetic acid and 1 ml of your alcohol sample. **Add one drop of concentrated sulfuric acid. (CAUTION!** *Use a dropper!*) Mix well and place in a boiling water bath for 15 minutes. Dilute with 2–3 ml of water, pour onto a watch glass or into a Petri dish, and, with a wafting motion of the hand over the mixture, determine if you can detect the fruity odor of ethyl acetate. If you are not familiar with the odor of ethyl acetate, your instructor will provide a sample.

Record all results of tests (B–D) on the report form.

Note 1. Freshly-prepared apple cider that is unpasteurized and that contains no preservatives must be used for this experiment. Frozen apple or grape juice concentrate containing no preservatives or additives (e.g., vitamin C) may be used successfully if fresh cider is not available.

PART II THE PREPARATION OF ACETIC ACID

The biosynthesis of acetic acid, unlike the preparation of ethyl alcohol, is an enzyme-catalyzed *oxidation* reaction that takes place when fruit juices are fermented in the presence of air. The alcohol which is produced first when apple cider ferments is further acted upon by various species of *Acetobacter* when the fermentation process is carried out in an open vessel. The reaction produces a vinegar in which acetic acid is the principal component and may be written in the following equation form:

$$CH_3CH_2OH + O_2(\text{air}) \xrightarrow{\textit{Acetobacter}} CH_3COOH + H_2O$$

The preparation of vinegar for use as a condiment and preservative is another chemical reaction little understood but practiced by humans for thousands of years. A number of fruit juices may be used as starting material for the preparation of vinegar, but that obtained from apples is the starting material for "brown" or cider vinegar — the kind we will prepare in the present experiment.

Procedure

Place 250 ml of freshly prepared apple cider in a 500-ml beaker or a glass jar, cover with a wire gauze, and set in a warm (25°C) place (Note 1). Allow the cider to ferment for 2–3 weeks, periodically stirring it to break up the slimy film (*mother of vinegar*) that forms on the surface and prevents free contact with air. When fermentation is complete, filter the mixture to remove the scum and residue that accumulate, then carry out the following experiments on the relatively clear filtrate.

(a) Determine the acidity (expressed as molarity of acetic acid) of a sample of the vinegar you have prepared by diluting a 25-ml portion with 50 ml of distilled water and titrating with a standardized sodium hydroxide solution. A base strength approximately 0.5 *M* is needed. Use phenolphthalein as an indicator (Note 2). The concentration of acetic acid in commercial vinegar usually is fixed at 5% (*ca.* 0.83 *M*). Next, take a pH measurement on a sample of the vinegar.

(b) Distill a 50-ml sample of your vinegar using a Bunsen burner as your heat source. Distill almost to dryness or until only 1–2 ml remains in the boiling flask (Note 3). Distillation will separate the volatile acetic acid from any solid acids that may be in solution. In addition to acetic acid, cider vinegar usually contains certain amounts of malic and tartaric acids. Again titrate a sample of the distillate as was done in Part (a), except that the dilution with water is unnecessary. Also, measure the pH of the distillate.

(c) Analyze a sample of commercial "white" vinegar by determining its molarity and pH values.

Record the results of your measurements in Parts (a–c) on your report form.

(d) From concentration measurements and pH values obtained in Parts (b) and (c), determine a value for the pK_a of acetic acid.

You may find the following relationships useful.

$$K_a = \frac{\text{Concentration of hydronium ion} \times \text{Concentration of acetate ion}}{\text{Concentration of undissociated acid}}$$

$$K_a = \frac{[H_3O^+][CH_3COO^-]}{[CH_3COOH]}$$

$$[H_3O^+] = [CH_3COO^-]$$

$$pH = \log \frac{1}{[H_3O^+]} = -\log [H_3O^+]$$

$$pK_a = -\log K_a$$

Note 1. See Note 1 under Part I.

Note 2. The yellow color of undiluted vinegar will mask the phenolphthalein color change at the end point.

Note 3. Frothing may present a problem. In this case, distillation should be carried out as judiciously and as long as possible.

TEXT REFERENCE

Report: 18

Chapter Pages

Section Desk NAME ______________________

ETHANOL

GLC ANALYSIS OF ETHANOL

Peak areas of prepared ethanol: C_2H_5OH __________; H_2O __________

Sample contained __________% ethanol; __________% water.
(Attach chromatogram)

OXIDATION OF ETHANOL

Explain the color changes that occurred when a sample of ethanol was treated with chromic acid. ______________________

____

Write a *balanced* equation for the reaction that took place by completing the following:

$$C_2H_5OH + \quad K_2Cr_2O_7 + \quad H_2SO_4 \rightarrow \quad CH_3COOH + \quad K_2SO_4 \quad + \quad Cr_2(SO_4)_3 + \quad H_2O$$

IODOFORM TEST

Describe the results of the iodoform test. ______________________

____

Write a balanced equation for the reaction that took place by completing the following:

$$C_2H_5OH + \quad NaOH + \quad I_2 \rightarrow \quad CHI_3 + \quad HCOO^-Na^+ + \quad NaI + \quad H_2O$$

ESTERIFICATION TEST

Describe the results of the esterification test. ______________________

____

ACETIC ACID IN VINEGAR

	Apple Cider Vinegar			Commercial White Vinegar		
	pH	$[CH_3COOH]$	pK_a	pH	$[CH_3COOH]$	pK_a
Average Values	2.70 2.95	0.823 0.727	5.06 5.76	2.50 2.48	0.850 0.839	4.13 4.88
Experimental Values						

K_a for acetic acid $= 1.78 \times 10^{-5}$; $pK_a = 4.75$.

Questions and Exercises

1. Champagne bottles are made of heavy glass and stoppers are wired tight. What is the reason for this?
2. Some beers are bottled before fermentation is complete, while others are bottled after fermentation is complete but are carbonated and bottled while very cold. Explain.
3. A natural wine usually has an alcoholic content no greater than 12–13% by volume. Why can the alcoholic content not exceed this?
4. King Arthur and his knights on occasion probably spent some time at the round table drinking mead. What is mead?
5. If a fractional distillation had been carried out a third or even a fourth time on our sample of fermented cider, could we have obtained pure ethanol? Explain.
6. Consult your text and describe the Weizmann fermentation. What products resulted from this historically and commercially important achievement?
7. What are the starting materials and methods for the production of ethanol other than by fermentation of carbohydrates? Write an equation to illustrate.
8. Consult your text or some other literature source to learn the meaning of *pyroligneous acid.* From what is it made? What is its composition?

Experiment 19

Time: 3½–4 hours

Preparation of Cyclohexene

The dehydration of alcohols with acids is a general laboratory method for preparing alkenes. Strong mineral acids, such as sulfuric or phosphoric acids, usually are used as catalysts. In this experiment cyclohexanol is dehydrated to produce cyclohexene according to the following equation.

$$\text{Cyclohexanol} \xrightarrow{H_3PO_4} \text{Cyclohexene} + H_2O$$

Cyclohexanol Cyclohexene

Cyclohexene (b.p. 83°), along with some water, is distilled from the reaction mixture as it is produced by careful heating of the mixture at or near 100°. Unreacted cyclohexanol (bp 161°) remains in the boiling flask to be further acted upon by the acid. By careful control of the temperature the reaction can be made to go to completion. The small amount of mineral acid that invariably appears along with the product is neutralized by washing the product with sodium carbonate. Traces of water are removed from the crude cyclohexene by drying the liquid over anhydrous calcium chloride (a salt that forms hydrates and removes water from many organic liquids).

Procedure

Set up a distillation apparatus as illustrated in Figure 3.1, supporting a 100-ml round-bottomed flask on an asbestos gauze. Inasmuch as the product is a very volatile and flammable hydrocarbon, use a special vented receiver of the type illustrated in Figure 19.1. If an electric heating mantle is used as your heat source, the hose leading from the side arm of the adapter may be eliminated. Cool the receiver in crushed ice and water.

Place 20 g (21 ml, 0.20 mole) of cyclohexanol (sp. gr. 0.96) and 5.0 ml of 85% phosphoric acid in the distilling flask (use the long-stemmed funnel) and mix thoroughly. Add a "Boileezer" or two to the mixture, start the water circulating gently through the condenser, and carefully heat the reaction flask with a small flame. Control the heating operation by holding the burner in

FIGURE 19.1 Receivers for highly volatile and flammable distillates vented *via* rubber tubing over side of laboratory desk.

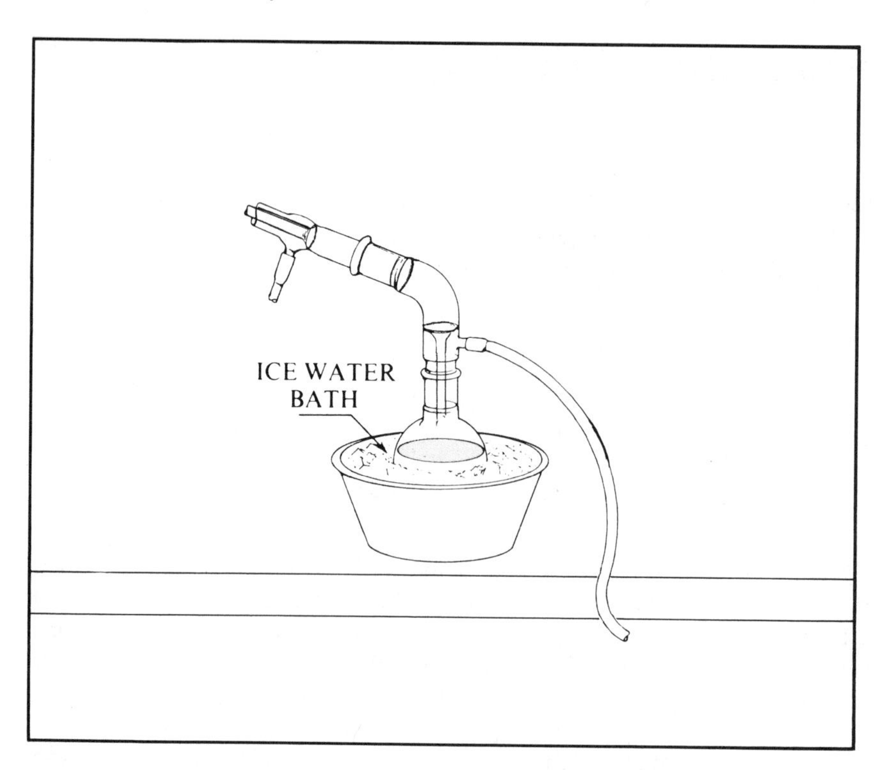

your hand, moving the burner as necessary to maintain the temperature. The temperature of the vapor condensing on the thermometer bulb should never register in excess of 100°. Continue the distillation until only 5–6 ml of residue remain in the boiling flask. Saturate the aqueous layer of the distillate with solid sodium chloride (approximately 0.5 g) and add enough 10% sodium carbonate solution to make the aqueous layer basic to litmus. Transfer the mixture to a separatory funnel and separate the lower aqueous layer from the hydrocarbon. Pour the cyclohexene through the neck of the funnel into a Erlenmeyer flask and add 1–2 g of anhydrous calcium chloride. During the drying interval (Note 1) clean, dry, and reassemble your distillation apparatus, but this time use a distilling flask of 50-ml capacity. Transfer the cyclohexene to the distilling flask by pouring it through a clean, dry, long-stemmed funnel fitted with a small cotton plug to retain the calcium chloride (Note 2). Add a "Boileezer" or two and distill, collecting the principal portion within a boiling range of 80–85°. Yield 10–12 g (79–80%). After weighing your product and calculating your yield (next section) use small samples of your product to perform the bromine and Baeyer tests described in Experiment 7, Part I-B. Save the remainder of your cyclohexene for use in Experiment 22.

THE CALCULATION OF YIELDS IN ORGANIC SYNTHESES

In all laboratory procedures that lead to the preparation of an organic compound, the amount of product of acceptable purity actually obtained is called a *yield.* The yield is dependent upon a number of factors, which include (1) the relative amount of each reactant (2) the temperature at which the reaction takes place (3) the possibility of side reactions giving unwanted byproducts, and (4) the mechanical losses inherent in the purification process. Therefore, very few organic syntheses result in a 100% yield or a complete conversion of all starting material to product.

The *percentage yield* is the ratio of product obtained to that theoretically possible × 100. If a number of reactants are involved in a synthesis the stoichiometry of which is that shown by $A + B + C \rightarrow D + E + F$, then the theoretical amount of product obtainable is limited by the number of moles of the reactant present in smallest amount. To calculate the percentage yield, find the number of moles of each reactant taken and the theoretical amount (in grams) of product obtainable from the limiting reactant. The percentage yield is then calculated from the following relationship:

$$\%\ \text{Yield} = \frac{\text{Weight of product actually obtained}}{\text{Moles of limiting reactant} \times \text{M.W. of product}} \times 100$$

In the report on the preparation of cyclohexene the theoretical yield will be calculated to illustrate the process. In subsequent preparations you will be expected to calculate the theoretical yields, using this experiment as a model.

Note 1. The drying process will be hastened if the flask is swirled occasionally. If there is insufficient time to complete the distillation in the same laboratory period as that in which the experiment was begun, the crude product may be left over calcium chloride until the next laboratory period.

Note 2. The cotton plug should be no larger than the tapered end of a sharpened pencil.

TEXT REFERENCE

Report: 19

Chapter Pages

Section Desk NAME ____________________

PREPARATION OF CYCLOHEXENE

Reaction equation

Cyclohexanol (M.W.100) $\xrightarrow{H_3PO_4}$ Cyclohexene (M.W.82) + H_2O

Amt. of reactant used[1] = 20.0 g. (21 ml)

Theoretical number of moles product obtainable $= \dfrac{20.0\text{ g}}{100\text{ g/mole}} = 0.20$

Amt. of product theoretically obtainable $= 0.20 \times 82 = 16.4$ g

Theoretical yield 16.4 g

Actual yield ________ g

$$\text{Percentage yield} = \frac{\text{Actual yield (g)}}{\text{Theoretical yield (g)}} \times 100 = \_\_\_\_\_\_$$

Results of tests with (a) Br_2—CH_2Cl_2 solution; (b) $KMnO_4$

(a) ____________________

(b) ____________________

[1]As cyclohexanol is the only reactant it is the limiting reactant.

Questions and Exercises

1. What alcohol would be the most logical choice as starting material for the preparation of 1-methylcyclohexene? Write the equation for the reaction. Explain why you think your choice is the most logical.
2. How much of the alcohol you chose in answering Question 1 would you require as starting material if you needed to prepare 25 g of 1-methylcyclohexene and if your actual percentage yield were going to be 75%?
3. What is the theoretical yield of 1,2-dibromocyclohexane for the reaction of 20 g of cyclohexene with 20 g of bromine? (*Hint:* Which is the limiting reactant?)
4. A reaction is carried out for which the stoichiometry is A + 2B + 3C → D + 2E + F. In this reaction 0.5 mole of A, 0.75 mole of B, and 1.0 mole of C are allowed to react. The yield of E is 0.2 mole. What is the percentage yield?
5. Consider the following reaction:

$$\underset{75\text{ ml}}{(CH_3)_2C{=}O} + \underset{8\text{ g}}{Mg} \rightarrow (CH_3)_2C(-O-)\text{—}C(CH_3)_2(-O-) \text{ (the two O atoms bridged by Mg)}$$

$$\xrightarrow{2\,H_2O} \underset{18\text{ g}}{(CH_3)_2C(OH)\text{—}C(CH_3)_2(OH)} + Mg(OH)_2$$

Density of acetone = 0.7908

Which is the limiting reagent in this reaction? What is the percentage yield?

Experiment 20

Time: $3\frac{1}{2}$–4 hours

Preparation of Alkyl Halides

Alkyl halides are among the most important chemical intermediates used by the organic chemist. Although a number of alkyl halides have valuable practical applications as solvents, pesticides, pharmaceuticals, etc., many others are prepared principally for use as intermediates or starting materials in the synthesis of more complex molecules. For this reason the organic chemist has developed a number of methods of synthesizing alkyl halides from readily available materials. This experiment illustrates the preparation of alkyl halides from alcohols, one of the most general methods available for making alkyl halides.

Replacement of the hydroxyl group of an alcohol by halogen may be carried out by treating the alcohol with a phosphorus halide, thionyl chloride, or with a halogen acid.

$$3\ \text{R—OH} + \underset{\text{Phosphorus trihalide}}{PX_3} \rightarrow 3\ \text{R—X} + \underset{\text{Phosphorous acid}}{H_3PO_3} \qquad (X = Br, Cl)$$

$$\text{R—OH} + \underset{\text{Phosphorus pentachloride}}{PCl_5} \rightarrow \text{R—Cl} + \underset{\text{Phosphorus oxychloride}}{POCl_3} + HCl$$

$$\text{R—OH} + \underset{\text{Thionyl chloride}}{SOCl_2} \rightarrow \text{R—Cl} + SO_2 + HCl$$

$$\text{R—OH} + HX \rightarrow \text{R—X} + H_2O \qquad (X = Br, Cl, I)$$

The choice of method to be used depends on many factors: cost, convenience, ease of purification, and, perhaps most important, the effect of the structure of the alcohol or the alkyl halide on the course of the reaction. In this experiment the replacement of the hydroxyl group is brought about through use of a halogen acid.

The replacement of a hydroxyl group by halogen using a halogen acid is an example of a nucleophilic substitution reaction in which one nucleophile is substituted for another. The most obvious exchange of nucleophilic groups, halide ion for hydroxide ion, is an experimentally impractical one because the equilibrium lies far to the left as written.

$$R{-}OH + X^- \rightleftharpoons R{-}X + OH^- \qquad \text{(Equilibrium far to left)}$$

Indeed the reverse reaction is used to prepare alcohols from alkyl halides. In general, it may be said that hydroxide ion is a poor leaving group. In the presence of a strong acid, however, the alcohol is protonated to some extent on the oxygen atom to form an oxonium ion, and a much more useful exchange can take place, halide ion for the water molecule. Small, neutral molecules, like the water molecule, are often excellent leaving groups.

$$R{-}OH + H^+ \rightleftharpoons R{-}\overset{+}{O}H_2$$

$$R{-}\overset{+}{O}H_2 + X^- \rightleftharpoons R{-}X + H_2O \qquad \text{(Position of equilibrium dependent on R—, X—, and conditions)}$$

In the acid-catalyzed reaction the position of the equilibrium is strongly dependent on the structure of the alkyl group (R—), the halogen (X—), and the conditions of the experiment, but generally conditions can be found in which the equilibrium is well toward the right.

The foregoing equations describe only the gross overall changes taking place during the reaction. The detailed description of the reaction pathway being followed, i.e., the reaction mechanism, indicates the reaction to be of the S_N1 or S_N2 type. (In order to better understand this experiment, it is suggested that you read the discussion of S_N1 and S_N2 reactions in your textbook.)

A. PREPARATION OF *n*-BUTYL BROMIDE

In the first part of this experiment *n*-butyl alcohol is treated with hydrobromic acid to produce *n*-butyl bromide according to the reaction equation below. The hydrobromic acid is prepared *in situ* from sodium bromide and sulfuric acid.

$$NaBr + H_2SO_4 \rightarrow HBr + NaHSO_4$$

$$\underset{n\text{-Butyl alcohol}}{CH_3CH_2CH_2CH_2OH} + HBr \rightarrow \underset{n\text{-Butyl bromide}}{CH_3CH_2CH_2CH_2Br} + H_2O$$

The conversion of *n*-butyl alcohol to *n*-butyl bromide is an excellent example of a reaction proceeding largely by an S_N2 mechanism.

$$CH_3CH_2CH_2CH_2{-}OH + H^+ \rightarrow CH_3CH_2CH_2CH_2{-}\overset{\oplus}{O}H{-}H$$

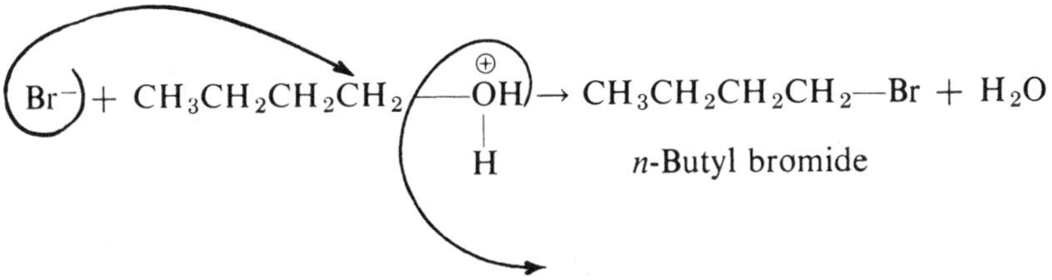

n-Butyl alcohol

$$Br^- + CH_3CH_2CH_2CH_2{-}\overset{\oplus}{O}H{-}H \rightarrow CH_3CH_2CH_2CH_2{-}Br + H_2O$$

n-Butyl bromide

However, when a primary alcohol is heated in the presence of a strong mineral acid, dehydration (Experiment 19) by an E1 or E2 mechanism also is possible. It is inevitable, therefore, that elimination occurs as a competing side reaction to produce some 1-butene.

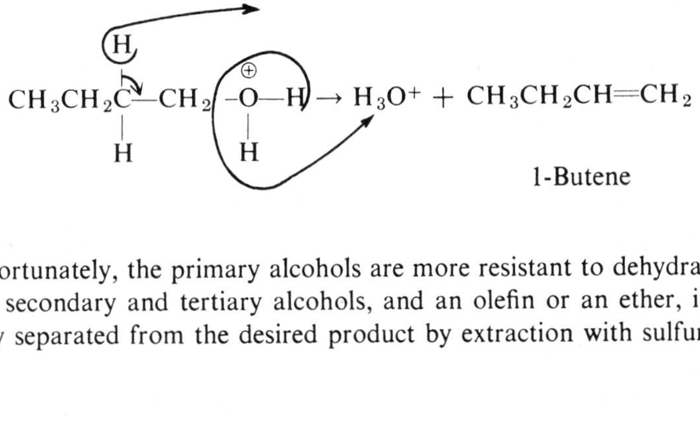

$$CH_3CH_2CH(H){-}CH_2{-}\overset{\oplus}{O}(H){-}H \rightarrow H_3O^+ + CH_3CH_2CH{=}CH_2$$

1-Butene

Fortunately, the primary alcohols are more resistant to dehydration than are the secondary and tertiary alcohols, and an olefin or an ether, if formed, is easily separated from the desired product by extraction with sulfuric acid.

Procedure

Place 31.0 g of sodium bromide, 35 ml of water, and 18.5 g (23 ml, 0.25 mole) of *n*-butyl alcohol in a 250-ml round bottomed flask. Cool the mixture in an ice water bath. Add 27 ml of concentrated sulfuric acid in 2- to 3-ml portions with thorough mixing (swirling) and cooling. Add a boiling chip, attach a reflux condenser (Fig. 2.3) and heat the mixture to boiling.

Maintain a vigorous reflux for 45 minutes. Discontinue heating and cool the reaction mixture by immersing the flask in a cold water bath. Remove the reflux condenser and fit the boiling flask with a connecting adapter and a condenser set for distillation (Fig. 3.1). Add a boiling chip and distill. The distillate will be composed of two layers—aqueous and organic. Continue distillation until the temperature reaches 110–115°. The upper oily layer in the distillation flask will have disappeared by this time and only water will distill over. You can verify this by collecting a few drops of distillate in a small (10 mm × 75 mm) test tube half-filled with water and noting whether the distillate contains droplets of oil or is composed only of water. Transfer the distillate to a separatory funnel and add 50 ml of water. Allow two layers to separate; withdraw and save the lower layer. Discard the upper water layer. Return the crude *n*-butyl bromide to your separatory funnel and successively wash it with 25 ml of cold concentrated sulfuric acid, 25 ml of water, and finally with 25 ml of 10% sodium carbonate solution. Allow ample time after each wash for a clean separation and be careful to save the proper layer. **CAUTION!** *Some effervescence might accompany the final wash; therefore, swirl the separatory funnel to obtain a preliminary mixing before you stopper and shake it.* Separate the "wet" *n*-butyl bromide into a 50-ml Erlenmeyer flask and add approximately 1–2 g of anhydrous calcium chloride. Swirl the flask occasionally to hasten the drying process. When the product is clear, transfer it by way of a long-stemmed funnel fitted with a small cotton plug into a 50-ml distilling flask, add a boiling chip, and distill. Collect the material boiling in the range of 98–102°. The boiling point of pure *n*-butyl bromide is 101°. Yield 20–25 g (58–74%). Record your yield on the report form and submit your product in a clean, labeled bottle to your laboratory instructor. Your product will be required in Experiment 21.

B. PREPARATION OF *tert*-BUTYL CHLORIDE

Although hydrobromic acid reacts with primary, secondary, and tertiary alcohols to form alkyl bromides fairly readily, hydrochloric acid reacts satisfactorily only with the more reactive alcohols: tertiary, allylic, and benzylic. In terms of the general discussion given above, this observation can be rationalized by stating that chloride ion is not as good a nucleophile as bromide ion. The lesser nucleophilicity of the chloride ion can be compensated for, in part, by employing a better leaving group. Thus, in some preparations of alkyl chlorides, zinc chloride (a Lewis acid) is added to change the nature of and increase the effectiveness of the leaving group. This is the basis of the Lucas Test [Experiment 10 (C)] for distinguishing between primary, secondary, and tertiary alcohols.

Tert-butyl alcohol reacts so readily with concentrated hydrochloric acid that zinc chloride need not be used in this experiment. The reason for this greater reactivity on the part of *tert*-butyl alcohol is explainable by the inductive and resonance stabilization of the intermediate carbocation which forms in the first stage of the reaction as indicated.

$$\text{I} \qquad CH_3-\overset{\overset{\large CH_3}{|}}{\underset{\underset{\large CH_3}{|}}{C}}-OH + HCl \rightarrow CH_3-\overset{\overset{\large CH_3}{|}}{\underset{\underset{\large CH_3}{|}}{C}}{\oplus} + Cl^- + H_2O$$

$$CH_3-\overset{\overset{\large CH_3}{|}}{\underset{\underset{\large CH_3}{|}}{C}}{\oplus} \leftrightarrow CH_3-\overset{\overset{\large H^{\oplus}CH_2}{\|}}{\underset{\underset{\large CH_3}{|}}{C}} \leftrightarrow H^{\oplus}CH_2{=}\overset{\overset{\large CH_3}{|}}{\underset{\underset{\large CH_3}{|}}{C}} \leftrightarrow CH_3-\overset{\overset{\large CH_3}{|}}{\underset{\underset{\large H\oplus CH_2}{\|}}{C}}$$

$$\text{II} \qquad (CH_3)_3C{\oplus} + Cl^- \rightarrow CH_3-\overset{\overset{\large CH_3}{|}}{\underset{\underset{\large CH_3}{|}}{C}}-Cl$$

The positive carbocation unites in the second stage of the reaction with the negative chloride ion. The preparation of *tert*-butyl chloride is an excellent illustration of a reaction proceeding principally by the S_N1 reaction mechanism.

Procedure

Place 65 ml of concentrated hydrochloric acid (sp. gr. 1.19, 36–38 per cent) in a 125-ml Erlenmeyer flask and cool it to 0–5° in an ice bath. Transfer the cooled acid to a 125-ml separatory funnel. Add 15 g (19 ml, 0.2 mole), of *tert*-butyl alcohol to the funnel (Note 1) and shake the mixture occasionally for an interval of 15 minutes. Release the internal pressure regularly during the shaking period. Allow the mixture to stand in the funnel until the layers have separated. Using the technique described in Experiment 5, draw off the lower layer through the stopcock (after testing it to make certain that it is the water layer) and discard it. Wash the crude product while still in the funnel by shaking it successively with 15 ml of water and with 15-ml portions of 5% sodium bicarbonate solution until the last traces of hydrochloric acid are neutralized. **CAUTION!** *Carbon dioxide is liberated in the neutralization step; swirl the liquids in the funnel for a few seconds before inserting the stopper and shaking.* Wash once more with 15 ml of water. After each washing withdraw the lower layer, test it, and discard it only if it is an aqueous layer. Transfer the crude *tert*-butyl chloride to a dry 50-ml Erlenmeyer flask and add 2 g of anhydrous calcium chloride. While the liquid is drying (Note 2), set up a distillation apparatus like that shown in Figure 3.1, p. 23, using a 50-ml distilling flask. The crude *tert*-butyl chloride, after drying over calcium chloride for the length of time it required you to assemble your distillation apparatus, may be filtered directly into your distillation flask. Use a long-stemmed funnel fitted with a *small* cotton plug to remove the particles of calcium chloride. Add 2–3 "Boileezers" or small pieces of clay plate and distill the product over a low flame. Collect the fraction boiling between 45–52° in either a weighed 50-ml Erlenmeyer or round-bottom flask (Note 3). Weigh your product and calculate your yield. Submit your sample to your instructor with your report.

Note 1. Tertiary butyl alcohol has a melting point of 25° and often is solidified in the bottle. It may be melted by placing the bottle in a warm water bath.

Note 2. The drying process may be hastened by periodic shaking of the alkyl halide and the drying agent.

Note 3. The major portion of product will distill at the lower end of this temperature range.

C. CLASSIFICATION TESTS FOR ALKYL HALIDES

In this part of the experiment the alkyl halides prepared in procedures (A) and (B) are tested for their reactivity toward two classification reagents, silver nitrate in ethanol solution and sodium iodide in acetone solution. In general, ethanolic silver nitrate tends to react by the S_N1 mechanism with alkyl halides to form an alkyl nitrate and an insoluble silver halide.

$$R—X + AgNO_3 \rightarrow R—ONO_2 + AgX$$

The order of reactivity of saturated, acyclic alkyl halides toward ethanolic silver nitrate is found to be:

$$\text{Tertiary} > \text{Secondary} > \text{Primary}$$

Tertiary halides generally react with immediate precipitation at room temperature. Primary and secondary halides react slowly, if at all, at room temperature but react readily at the boiling point of ethanol to give a precipitate. In general, sodium iodide in acetone tends to react with alkyl halides by the S_N2 mechanism to form the alkyl iodide and the insoluble (in acetone) sodium bromide or sodium chloride.

$$R—X + NaI \rightarrow R—I + NaX$$

The order of reactivity of saturated, acyclic alkyl halides toward sodium iodide in acetone is found to be:

$$\text{Primary} > \text{Secondary} > \text{Tertiary}$$

With this reagent, primary bromides give a precipitate of sodium bromide within 3 minutes at room temperature, whereas primary chlorides must be heated to 50° to bring about a reaction. Secondary chlorides and secondary and tertiary bromides react at 50°, but tertiary chlorides react too slowly to give a positive test.

By employing both of these reagents, valuable information about the structure of an alkyl halide can be obtained.

Procedure

Place 2 ml of a 2% ethanolic silver nitrate solution in each of two test tubes. Add one drop of your *n*-butyl bromide to one test tube and one drop of your *tert*-butyl chloride to the other test tube. Note and record whether or not a precipitate is formed in either tube. If no reaction is observed after 5 minutes, heat the solution to boiling in the steam bath. (**CAUTION!** *Ethanol is flammable.*)

Place 1 ml of the acetone solution of sodium iodide in each of two *dry* test tubes. Add one drop of your *n*-butyl bromide to one test tube and one drop of your *tert*-butyl chloride to the other test tube. Shake each tube to mix the contents, and allow the solutions to stand for 3 minutes. Note and record whether or not a precipitate is formed in either tube. If no precipitate has formed, place that tube in a beaker of water heated to 50°. (**CAUTION!** *Acetone is flammable; turn off burner before placing the test tube in the water bath.*) Allow the solution(s) to stand in the hot water for 6 minutes. Remove the test tube(s) and cool to room temperature. Note and record whether or not a precipitate is formed.

If the results obtained in either the silver nitrate or sodium iodide tests appear to be inconsistent with the general statements made in the introduction to this part of the experiment, repeat the tests with *n*-butyl bromide and *t*-butyl chloride taken from the reagent shelf.

TEXT REFERENCE

Report: 20

Chapter Pages

Section Desk NAME

PREPARATION OF *n*-BUTYL BROMIDE

Reaction equation $CH_3CH_2CH_2CH_2OH + HBr \rightarrow CH_3CH_2CH_2CH_2Br + H_2O$

Quantities	18.5 g (23 ml)	________
Mol. Wt.	________	________
Moles	________	________
Equivalents	________	________

Theoretical yield ________ g

Actual yield ________ g

Percentage yield ________

PREPARATION OF *tert*-BUTYL CHLORIDE

Reaction equation

$$CH_3-\underset{CH_3}{\overset{CH_3}{\underset{|}{\overset{|}{C}}}}-OH + HCl \rightarrow CH_3-\underset{CH_3}{\overset{CH_3}{\underset{|}{\overset{|}{C}}}}-Cl + H_2O$$

Quantities	15 g (19 ml)	________
Mol. Wt.	________	________
Moles	________	________
Equivalents	________	________

Theoretical yield ________ g

Actual yield ________ g

Percentage yield ________

CLASSIFICATION TESTS

Record the appearance of a precipitate with a plus (+) and the absence of a precipitate with a minus (−).

Compound	Silver nitrate test		Sodium iodide test	
	Unheated	Heated	Unheated	Heated
Your *n*-Butyl Bromide				
Reagent *n*-Butyl Bromide				
Your *tert*-Butyl Chloride				
Reagent *tert*-Butyl Chloride				

Questions and Exercises

1. Write balanced equations to show how a sulfuric acid wash extracts the by-products that inevitably result when *n*-butyl bromide is prepared by the method used in this experiment.
2. What products would you predict if isopropyl alcohol rather than *n*-butyl alcohol had been the starting material in this experiment?
3. Devise a procedure for the preparation of *n*-butyl bromide other than the one used in this experiment but one which begins with the same reagents.
4. What impurity could cause either *n*-butyl bromide or *tert*-butyl chloride, prepared as in this experiment, to give a false positive test with the two classification reagents?
5. Outline a procedure for the preparation of *sec*-butyl chloride. Include steps for the removal of impurities.

Experiment 21

Time: $3\frac{1}{2}$–4 hours

The Grignard Reaction: Preparation of 2-Methyl-2-Hexanol

The preparation of a Grignard reagent is a very rewarding experience for every organic chemistry student. It is a reaction which frequently tries one's patience, for Grignard reagents do not always form immediately, and often require some coaxing. It is especially important that glassware and all reagents be completely dry, because moisture will destroy the Grignard reagent according to the following equation:

$$RMgX + H_2O \rightarrow RH + Mg(OH)X$$

It is advantageous to dry glassware in a drying oven overnight before use, if possible. All reagents must be anhydrous. Once the Grignard reagent is prepared it should be used without delay and with a minimum exposure to air.

Procedure

Arrange on *only one* ring stand an assembly like that illustrated in Figure 21.1 using a *dry*, 250-ml round-bottomed flask (Note 1). Place in the flask 2.4 g (0.10 gram-atom) of magnesium turnings and 15 ml of *anhydrous* ether. Weigh in a clean, *dry*, 50-ml Erlenmeyer flask 13.7 g (10.75 ml, 0.10 mole) of previously prepared *n*-butyl bromide and transfer it to the separatory funnel. Fill the 50-ml Erlenmeyer flask with *anhydrous* ether and add it also to the dropping funnel. Swirl the funnel to insure complete mixing of the *n*-butyl bromide and ether. Add in one portion approximately 15 ml of the ether solution of *n*-butyl bromide from the funnel to the magnesium turnings. Reaction usually begins within a few minutes and is accompanied by a spontaneous change in appearance from clear to an opalescent white and a gentle boiling of the ether. Add the remainder of the reagent from the separatory funnel dropwise to maintain a fairly rapid reflux. If the reaction becomes too lively, stop addition of the reagent and immerse the round-bottomed flask in a cold water bath. Too rapid reflux will cause ether vapor to escape from the top of the condenser. If the reaction does not begin spontaneously, warm the flask by immersing it in a warm (50°) water bath. It may be necessary to gently crush the magnesium turnings with a large glass stirring rod slightly flattened on one end. This step insures exposure of bare magnesium metal to the alkyl halide and usually initiates reaction immediately. If this step is necessary you will need the help of your instructor or that of another student to lift the Claisen adapter with its attached dropping funnel and condenser clear of the reaction flask. Be sure to support the round-bottom flask in your hand and make certain that there are no open flames nearby.

Once the reaction is in progress, the mixture becomes progressively darker. When refluxing subsides and nearly all of the magnesium has been consumed, immerse the reaction flask in a 50° water bath and allow the mixture to reflux for an additional 15–20 minutes. At the end of this time replace the warm water bath with one containing cold water and cool the mixture. While the mixture is cooling prepare a solution of 5.8 g (7.4 ml, 0.10 mole) of dry acetone (Note 2) in 15 ml of anhydrous ether. Transfer the acetone-ether solution to the separatory funnel (be certain the stopcock is closed) and add it dropwise to the now cold Grignard reagent, swirling the flask frequently by giving a rotatory motion to the ring stand assembly. The reaction is very vigorous and good mixing is essential to insure proper cooling. After all acetone has been added allow the reaction mixture to stand at room temperature for 20–30 minutes (Note 3). Hydrolyze the Grignard addition compound by pouring it very slowly with stirring onto a cold mixture of 75 ml of saturated ammonium chloride and 25 g of crushed ice contained in a 250-ml beaker. Avoid transferring any unreacted magnesium. Rinse the reaction flask with a little ordinary ether and add the rinsings to the hydrolysis mixture. Stir until all solid material in the ether layer has dissolved. Transfer the mixture to a separatory funnel and separate. Extract the aqueous layer with a 25-ml portion of ordinary ether and add this ether extract to the separated ether solution. Discard the aqueous layer. Wash the ether once with 25 ml of 10% sodium carbonate solution and a second time with a 25-ml portion of saturated sodium chloride solution. Discard all aqueous washes. Dry the ether over anhydrous magnesium sulfate until the next laboratory period. Transfer the dried ether solution via a funnel fitted with a small cotton plug to a fractional distillation assembly (Fig. 4.1) and remove the ether using a hot water bath. After all ether has been removed, the residue may be transferred to a smaller distillation assembly and the product distilled by heating with an electric mantle or over a very small flame. Collect as your desired product all material distilling in a temperature range of 137–143°. The boiling point of 2-methyl-2-hexanol is reported to be 143°. Yield, approximately 7.0 g (60%).

Note 1. The glassware must be dry. If it has not been dried in the oven overnight, it may be advisable to flame-dry the flask with a burner flame before assembling the apparatus. **CAUTION!** *Ether is extremely flammable!* Do not light a burner if you or your neighbor are using ether.

Note 2. Acetone should be dried over anhydrous magnesium sulfate at least overnight.

Note 3. The reaction flask may be stoppered at this stage and the mixture worked up the following laboratory period.

FIGURE 21.1 Assembly for the preparation of a Grignard reagent. Notice that the system is open to the atmosphere.

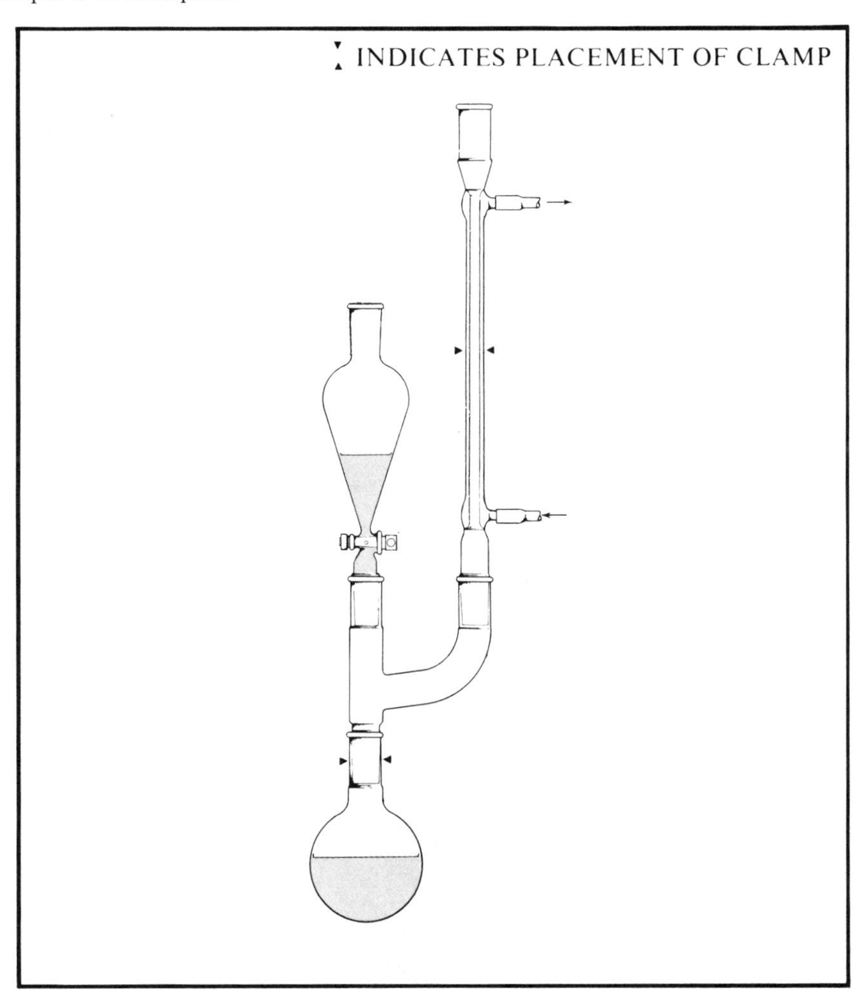

TEXT REFERENCE

Report: 21

Chapter Pages

Section Desk NAME

THE GRIGNARD REACTION: PREPARATION OF 2-METHYL-2-HEXANOL

Reaction equation

$$\underset{n\text{-Butyl bromide}}{CH_3(CH_2)_2CH_2Br} \xrightarrow[3.\ H_3O^+]{1.\ Mg,\ ether \quad 2.\ (CH_3)_2C{=}O} \underset{\text{2-Methyl-2-hexanol}}{CH_3(CH_2)_3\overset{\overset{CH_3}{|}}{\underset{\underset{OH}{|}}{C}}{-}CH_3}$$

Quantities 13.7 g (10.75 ml) ________

Mol. Wt. ________ ________

Moles ________ ________

Theoretical yield________g

Actual yield________g

Percentage yield________

Questions and Exercises

1. Some grades of ether may be anhydrous yet contain traces of ethyl alcohol. Would the presence of this impurity interfere in the preparation of a Grignard reagent? Explain.
2. Why is anhydrous magnesium sulfate rather than anhydrous potassium carbonate the preferred drying agent for acetone?
3. An attempted preparation of 2-methyl-2-hexanol appeared to proceed normally, but in the final distillation, after the ether had been removed, the only product recovered passed over at 85–90°. What product could this have been?
4. Write equations for two different preparations of 1-hexanol using a Grignard reagent.
5. Why are the Grignard preparations sometimes slow to start even when reagents are anhydrous and glassware dry?
6. The solvent of choice in a Grignard preparation is an anhydrous ether. Why could a Grignard not be prepared using an excess of the halogen compound as the solvent?
7. Using any readily available reagents and a Grignard reaction, show how the following alcohols might be prepared: (a) 3-methyl-1-butanol; (b) 2-phenyl-2-butanol; (c) 2,3,3-trimethyl-2-butanol.

Experiment 22

Time: $1\frac{1}{2}$–2 hours

Chromic Acid Oxidation: Preparation of Adipic Acid

Chromic acid, H_2CrO_4, is one of the most potent oxidizing agents used by the organic chemist. This reagent is generated by the treatment of either sodium dichromate or potassium dichromate with sulfuric acid according to the following equation:

$$Cr_2O_7^{2-} + 2\,H^+ \rightarrow 2\,CrO_3 + H_2O$$

$$CrO_3 + H_2O \rightarrow 2\,H^+ + CrO_4^{2-}$$

At room temperature chromic acid is capable of smoothly oxidizing a primary alcohol to a carboxylic acid and a secondary alcohol to a ketone and serves as a diagnostic test useful in the classification of alcohols (Exp. 10-D). Under more vigorous conditions chromic acid is capable of oxidizing an alkylated benzene to benzoic acid, effectively cleaving carbon-carbon bonds in the side chain down to the ring-attached carbon. The latter is oxidized to the carboxyl group, which is the highest oxidation state (3^+) permitted for one carbon atom attached to another.

Ring-opening oxidations leading to the formation of adipic acid may be accomplished using cyclohexanol or cyclohexanone as starting material. However, in both of these reactions the carbon atoms attacked are already bonded to oxygen. In the present experiment we will employ chromic acid to cleave the cyclohexene ring at its most vulnerable site—the carbon-carbon double bond.

Chromic acid oxidations are quite exothermic and careful control of the temperature must be maintained by the proper rate of addition of the oxidizing reagent. Should the reaction become too vigorous, fragmentation of the carbon chain and random oxidation may result.

Procedure

CAUTION! *A potassium dichromate–sulfuric acid solution is an extremely corrosive reagent. A rubberized apron or some protective garment should be worn while carrying out the following procedure.*

Place 25 g of ice in a 500-ml Erlenmeyer flask and cautiously add 30 ml of concentrated sulfuric acid. Mix by swirling the flask, then cool the acid solution in a cold water bath. Clamp the flask in place. While the acid solution is cooling to room temperature, prepare in a 125-ml Erlenmeyer flask a second solution of 9 g of potassium dichromate ($K_2Cr_2O_7$) in 35 ml of water to which

10 ml of concentrated sulfuric acid have been added. Swirl until the dichromate has dissolved. Transfer the dichromate solution to a small separatory funnel and support the funnel on a ring stand. Remove the flask containing the sulfuric acid solution from the cooling bath and add 4 ml (3.25 g, 0.04 mole) of cyclohexene. Swirl to dissolve as much of the hydrocarbon as possible. Add the potassium dichromate solution dropwise in approximately 1-ml increments, with swirling, to the cyclohexene–sulfuric acid mixture (Note 1). Keep the temperature of the reaction mixture at approximately 50°. The flask should feel quite warm but not too hot to hold. The addition should require 25–35 minutes. After all the dichromate solution has been added, the mixture should be a dark greenish blue (Note 2). Support the flask on a steam bath or in a hot-water bath (85–90°) and warm for 30 minutes (HOOD! NO FLAMES!) Periodically swirl the flask while heating. This step will help bring the reaction to completion and remove unreacted cyclohexene by evaporation. Next, pour the reaction mixture into a 400-ml beaker and allow the beaker to stand in an ice-water bath until the temperature is below 5°. Stir occasionally with a glass rod (not with your thermometer) and periodically scratch the walls of the beaker. When the temperature has reached 5°, small crystals of adipic acid should be seen on the surface of the liquid. Collect the adipic acid in a small Büchner funnel, and wash with only a small volume (2–3 ml) of ice water. The crystals of adipic acid will be fairly pure as collected on the filter and will give a sharp melting point. Should your product have a slight greenish color, it may be recrystallized from a small volume of hot water. However, before recrystallizing your sample of adipic acid, use the values from the following solubility table to draw a solubility–temperature curve and from this curve determine the volume of water you must employ in order to recover a maximum amount of product.

TABLE 22.1

SOLUBILITY OF ADIPIC ACID

Temperature, °C	g/100 g H_2O
15	1.44
40	5.12
50	9.24
60	17.6
70	34.1
100	100

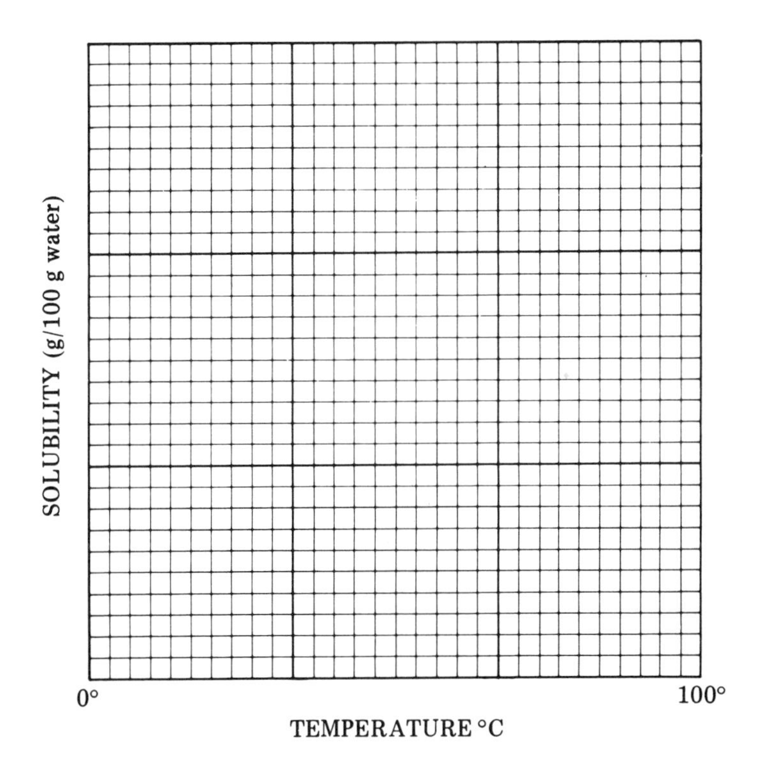

Note 1. Reaction is immediate, as will be indicated by a color change and a temperature rise. If swirling is done above a white buret base or a piece of paper, reaction progress may be noted by a sequential color change in the liquid film from orange to green to blue.

Note 2. The procedure may be interrupted at this stage and the reaction mixture allowed to stand overnight or until the following laboratory period. Actually, an improvement in the yield of adipic acid results from this additional reaction time.

TEXT REFERENCE

Report: 22

Chapter Pages

Section Desk NAME

PREPARATION OF ADIPIC ACID

$$3\ \text{(Cyclohexene)} + 4\ K_2Cr_2O_7 + 16\ H_2SO_4 \longrightarrow 3\ HOOC{-}(CH_2)_4{-}COOH\ \text{(Adipic acid)} + 4\ K_2SO_4 + 4\ Cr_2(SO_4)_3 + 16\ H_2O$$

Cyclohexene

Adipic acid

Quantities	3.25 g	9.0 g	________
Mol. Wt.	________	________	________
Moles	________	________	________

Theoretical yield______g

Actual yield______g

Percentage yield________

mp ______°C

Questions and Exercises

1. From the reaction equation for the oxidation of cyclohexene, how many equivalents of $K_2Cr_2O_7$ were taken per equivalent of cyclohexene?
2. What would the oxidation product have been if cyclohexene had been treated with ozone, followed by reductive hydrolysis?
3. A pseudoperspective drawing of the cyclohexene structure will show four of the ring carbon atoms in a tetrahedral bonding arrangement and four of the ring carbons lying in the same plane. Illustrate.
4. Write separate balanced oxidation-reduction reactions showing all products that result when the three isomeric butenes are treated with potassium dichromate and sulfuric acid.
5. If the solubility of adipic acid at 0°C is 0.80 g per 100 g water, and the total yield of adipic acid obtained from the oxidation of a sample of cyclohexene amounted to only 0.87 g, what is the maximum volume of water that should be used to recrystallize the sample and recover 0.80 g of pure product?
6. Employing a different set of reagents for each reaction, show how the following compounds can be oxidized into the products shown:

   (a) toluene → benzoic acid
   (b) 2-butanone → propanoic acid
   (c) glucose → gluconic acid
   (d) 2,3-butanediol → acetaldehyde

Experiment 23

Time: $3\frac{1}{2} + 3\frac{1}{2} + 2$ *hours*

Acid Derivatives: Acyl Chlorides, Anhydrides, and Amides

Usually acyl chlorides and acid anhydrides are prepared as intermediates in a reaction sequence, rather than as end products. Amides are often prepared as derivatives of the acids as part of the identification or characterization of the acids, although amides are also important intermediates in a number of reaction sequences (see Experiment 32). Experiment 23 illustrates general methods commonly used for the preparation of all three classes of compounds. A general method for the preparation of an acyl chloride and an amide as a part of the characterization process is illustrated in Part I. The preparations of phenylacetyl chloride and phenylacetamide as intermediates in a reaction sequence are illustrated in Part II. Benzoic anhydride is prepared and purified in Part III.[1]

Acyl chlorides are generally prepared by treating carboxylic acids with phosphorus trichloride, phosphorus pentachloride, or thionyl chloride.

$$3\,R{-}C(=O){-}OH + PCl_3 \rightarrow 3\,R{-}C(=O){-}Cl + H_3PO_3$$

$$R{-}C(=O){-}OH + PCl_5 \rightarrow R{-}C(=O){-}Cl + POCl_3 + HCl$$

$$R{-}C(=O){-}OH + SOCl_2 \rightarrow R{-}C(=O){-}Cl + SO_2 + HCl$$

Each of these reagents has its advantages and disadvantages and no one of them can be used for all conversions of acids to acid chlorides. However, many organic chemists prefer the use of thionyl chloride, when it is applicable,

[1]**Note to the instructor.** All of the acids and acyl chlorides used in this experiment are available commercially at a reasonable price. Therefore, the preparation of the acyl chlorides can be avoided if facilities do not permit them to be made conveniently. Most instructors will not want to assign both Parts I and II. Part III can be begun while Parts I or II are in progress.

because it is relatively simple to work with, is easily removed by distillation if employed in excess, and gives only volatile gases as by-products. Thionyl chloride is less suitable than phosphorus trichloride for the preparation of low-boiling acyl halides because of difficulties in separating the product from the reagent and removing the dissolved gaseous by-products.

The set-up and procedure for Parts I and II are almost identical, except for reaction temperature and work-up. Usually the reaction is conducted at the boiling point of thionyl chloride (75–76°) for a short period of time; phenylacetyl chloride may give undesired by-products at higher temperatures. Also in many examples a small amount of *dry* cyclohexane is added at the end of the reaction before the evaporation of unreacted thionyl chloride and other volatile materials to help sweep out these substances; then the acyl chloride is purified by vacuum distillation (Experiment 25). For simplicity these useful additional steps have been omitted in this experiment.

Amides can be prepared from acids (or acid derivatives) in a variety of ways (see discussion section of Experiment 26); however, one of the most general methods available is based on the reaction of an acyl chloride with ammonia or an amine.

$$R{-}C(=O)Cl + 2\,NH_3 \rightarrow R{-}C(=O)NH_2 + NH_4Cl$$

$$R{-}C(=O)Cl + 2\,NHR'_2 \rightarrow R{-}C(=O)NR'_2 + NH_2R'_2Cl$$

In general two equivalents of ammonia or amine are required, one being consumed in the neutralization of the hydrogen chloride liberated in the reaction. Reactions of acyl chlorides with amines are carried out by adding a solution of the acid chloride in methylene chloride to two equivalents of the amine in the same solvent. The insoluble amine hydrochloride generally precipitates from solution and may be removed by suction filtration. The amide is recovered by evaporation of the solvent. Reactions of acyl chlorides with ammonia can be carried out in the same fashion, substituting acetonitrile for methylene chloride; however, this procedure is not convenient in the beginning organic laboratory. Therefore, in this experiment the crude acyl chloride prepared in procedure (A) is added to an excess of cold, concentrated ammonium hydroxide. The insoluble amide precipitates from solution. Preparations of amides in this fashion, although quite common, suffer from the disadvantage that part of the acyl chloride is hydrolyzed back to the acid, which dissolves in the basic solution. Therefore, yields by this procedure are often rather low, although the product is generally fairly pure.

PART I PREPARATION OF *trans*-CINNAMAMIDE

The following procedure is very general and is widely used for the preparation of amides from acids *via* the acyl chloride.

$$C_6H_5\text{—CH=CH—C(=O)OH} \xrightarrow{SOCl_2} C_6H_5\text{—CH=CH—C(=O)Cl} \xrightarrow{NH_4OH} C_6H_5\text{—CH=CH—C(=O)NH}_2$$

Cinnamic Acid — Cinnamoyl Chloride — Cinnamamide

Procedure

A. PREPARATION OF *trans*-CINNAMOYL CHLORIDE

CAUTION! *Thionyl chloride is a low-boiling liquid (bp 75–76°) which is a strong irritant to the skin, eyes, and mucous membranes. It reacts vigorously with water to liberate hydrogen chloride and sulfur dioxide, which are also strong irritants. If the reagent comes into contact with the skin or clothing or is spilled on the bench, it should be washed off with copius quantities of water followed by dilute aqueous sodium bicarbonate. Thionyl chloride also attacks rubber; therefore, it is best handled in all-glass equipment with silicone grease on the ground joints. It should be handled only in a good fume hood* (Note 1).

In the hood set up a reflux apparatus like that shown in Figure 2.3, using your smallest round-bottomed flask (25–100 ml). Temporarily remove the round-bottomed flask, weigh it, and place in it 3.0 g of *trans*-cinnamic acid, 2.5 g (1.5 ml) of pure thionyl chloride (Note 2), and two or three "Boileezers". Immediately reassemble the apparatus. Heat the flask *gently* on the steam bath or over a very low flame for 45 minutes. The evolution of gas should begin almost immediately, and the acid should dissolve in the thionyl chloride within the first 15 minutes of the heating period.

Allow the apparatus to cool. Replace the reflux condenser with a rubber stopper fitted with a short length of glass tubing bent to a right angle. Connect the tubing to a trap of the type shown in Figure 2.4 by means of a short length of suction hose. Connect the trap to the water aspirator and *cautiously* reduce the pressure in the system to evaporate as much of the volatile material present as possible. When it appears that all of the easily removed volatile materials have been evaporated (about 5–10 minutes), release the vacuum, weigh the flask and crude product, and stopper the flask with the drying tube taken from the reflux condenser. The crude acid chloride should be used immediately for the preparation of the amide.

Note 1. Use of a cork to attach the drying tube is acceptable. If the hood facilities are not adequate to serve the whole laboratory section, the experiment can be run on the open bench using a gas trap of one of the types (a) or (b) of Figure 31.1.

Note 2. The best available grade of thionyl chloride should be used (e.g., Matheson Coleman and Bell reagent grade).

Procedure

B. PREPARATION OF *trans*-CINNAMAMIDE

Place 30 ml of concentrated ammonium hydroxide in a 125-ml Erlenmeyer flask, stopper loosely with a clean rubber stopper, and chill the solution thoroughly in the ice bath. When the solution is ice cold, slowly add the *trans*-cinnamoyl chloride prepared in procedure (A) of this experiment (Note 1), using a medicine dropper and stirring the mixture with each addition. There may be some sputtering and spattering as each drop reacts vigorously with the ammonium hydroxide. After the addition is complete, chill the mixture thoroughly (with occasional stirring) and collect the crude product on the Büchner funnel by suction filtration. Recrystallize the product from methanol (**CAUTION!** *Flammable*) by dissolving it in the minimum amount of hot methanol and adding water as required (Note 2). Collect the product by suction filtration, dry it thoroughly in the air on a clean piece of filter paper, weigh it, and determine its melting point. Yield, 1.4–2.4 g, mp 146–147°.

Note 1. If procedure (A) was not assigned or if the *trans*-cinnamoyl chloride prepared appears to be of low quality, 3.0 g of the commercially available acyl chloride may be substituted.

Note 2. Because many amides are quite soluble in hot methanol, it may be necessary to add water dropwise to the hot solution just to the point where the solution becomes faintly turbid. If necessary the hot solution may be clarified by addition of a very small amount of methanol. Chill and collect the product. The use of excess pure solvent is sometimes necessary when the minimum volume of solvent is so small as to make further manipulations difficult. Addition of a second poorer solvent then reduces the solubility to an acceptable level for a reasonable recovery of product. Avoid adding too large an initial excess of pure solvent, for it may then be difficult to get an acceptable recovery even by adding a second poorer solvent.

PART II PREPARATION OF PHENYLACETAMIDE

The following procedure, like that in Part I, is very general; however, because of the lower reaction temperature and longer reaction time it is usually employed where the higher temperature might lead to undesirable side products.

$$C_6H_5{-}CH_2{-}C(=O){-}OH \xrightarrow{SOCl_2} C_6H_5{-}CH_2{-}C(=O){-}Cl \xrightarrow{NH_4OH} C_6H_5{-}CH_2{-}C(=O){-}NH_2$$

Phenylacetic acid — Phenylacetyl chloride — Phenylacetamide

Procedure

A. PREPARATION OF PHENYLACETYL CHLORIDE

CAUTION! *Thionyl chloride is a low-boiling liquid (bp 75–76°) which is a strong irritant to the skin, eyes, and mucous membranes. It reacts vigorously with water to liberate hydrogen chloride and sulfur dioxide, which are also strong irritants. If the reagent comes into contact with the skin or clothing or is spilled on the bench, it should be washed off with copius quantities of water followed by dilute aqueous sodium bicarbonate. Thionyl chloride also attacks rubber and cork; therefore, it is best handled in all-glass equipment with silicone grease on the ground joints. It should be handled only in a good fume hood.*

In the hood, set up a reflux apparatus like that shown in Figure 2.3, except that the top of the reflux condenser is fitted with a calcium chloride tube (Note 1), using your smallest round-bottomed flask (25–100 ml). (No source of heat is required for this experiment). Temporarily remove the round-bottomed flask, weigh it, and place in it 2.8 g of phenylacetic acid (Note 2), 2.5 g (1.5 ml) of pure thionyl chloride (Note 3), and two or three "Boileezers" (Note 4). Immediately reassemble the apparatus. The evolution of gas should begin almost immediately and the contents of the flask may become warm from the heat of reaction. The mixture should be swirled carefully several times during the first two hours of reaction and then allowed to stand until the next laboratory period (Note 5). Be sure to label your set-up with your name to facilitate its identification.

Replace the reflux condenser with a rubber stopper fitted with a short length of glass tubing bent to a right angle. Connect the tubing to a trap of the type shown in Figure 2.4 by means of a short length of suction hose. Connect the trap to the water aspirator and *cautiously* reduce the pressure to evaporate as much of the volatile material present as possible. When it appears that all of the easily removed volatile materials have been evaporated (about 5–10 minutes), release the vacuum, weigh the flask and crude product, and stopper the flask with the drying tube taken from the reflux condenser. The crude acid chloride should be used immediately in procedure (B).

Note 1. Use of a cork to attach the drying tube is acceptable. If the hood facilities are not adequate to serve the whole laboratory section, the experiment can be started on the open bench using a gas trap of one of the types (a) or (b) of Figure 31.1. The set-up can be transferred to the hood at end of the period (3–4 hours) without water in the reflux condenser. In any of these set-ups the drying tube should be used during the time between laboratory periods.

Note 2. The lower aliphatic acids (butyric through capric) are liquids or low-melting solids which are relatively safe to work with but which have distinctly unpleasant odors. Phenylacetic acid is also an aliphatic acid with an aromatic substituent (phenyl) on the α-carbon atom. It, too, has a very unpleasant odor. Therefore, extreme care must be taken to keep these materials off of the skin, clothing, books, and the laboratory bench. It is advisable to use plastic gloves of the throw-away type in working with these acids. Furthermore, all glassware coming into contact with these acids should be washed immediately after use or dissembly. Very dilute (5%) aqueous sodium hydroxide followed by copius quantitities of water or, for gummy or tarry residues, acetone followed by water will remove these materials very effectively from glassware or the laboratory bench. If any of these acids come in contact with the skin, wash them off immediately with soap and water. All of these substances are absorbed to some extent by the skin and cannot be completely washed off with soap and water. If a good scrubbing is employed, any residual odor can be masked fairly well for several hours with a good spray deodorant. Usually the last traces of odor will be gone in 2–3 days.

The acid chlorides of these acids react with moisture to reform the acids; therefore, the acid chlorides must be treated with the same care.

Note 3. The best available grade of thionyl chloride should be used (e.g., Matheson Coleman and Bell reagent grade).

Note 4. Boiling stones are essential to prevent bumping.

Note 5. The water passing through the reflux condenser may be turned off at the end of the laboratory period (3–4 hours). If the standing period is less than several days, it may be necessary to complete the reaction by warming the flask with a warm water bath heated to 40–45° until gas evolution ceases. Do *not* heat the mixture more strongly than this.

Procedure

B. PREPARATION OF PHENYLACETAMIDE

Place 30 ml of concentrated ammonium hydroxide in a 125-ml Erlenmeyer flask and chill the solution thoroughly in the ice bath. When the solution is ice cold, slowly add the crude phenylacetyl chloride prepared in procedure (A) of this experiment (Note 1), using a medicine dropper and stirring the mixture with each addition. There may be some sputtering and spattering as each drop reacts with the ammonium hydroxide. After the addition is complete, rinse the flask that contained the acid chloride with 5 ml of concentrated ammonium hydroxide and add to the Erlenmeyer flask. Chill the mixture thoroughly (with occasional stirring) and collect the crude product on the Büchner funnel by suction filtration. Recrystallize the product from methanol (**CAUTION!** *Flammable*) by dissolving it in the minimum amount of hot methanol and adding water as required. (Note 2) Collect the product by suction filtration, dry it thoroughly in the air on a clean piece of filter paper, weigh it, and determine its melting point. Yield, 1.4–1.8 g; mp 157–158°.

Note 1. If procedure (A) was not assigned of if the phenylacetyl chloride prepared appears to be of low quality, 3.0 g of the commercially available phenylacetyl chloride may be substituted.

Note 2. Because the product is quite soluble in hot methanol, it will be necessary to add water dropwise to the hot solution just to the point where the solution becomes faintly turbid. If necessary the hot solution may be clarified by addition of a very small amount of methanol. Chill and collect the product. The use of excess pure solvent is sometimes necessary when the minimum volume of solvent is so small as to make further manipulations difficult. Addition of a second poorer solvent then reduces the solubility to an acceptable level for a reasonable recovery of product. Avoid adding too large an excess of pure solvent, for it may then be difficult to get an acceptable recovery even by adding a second poorer solvent.

PART III PREPARATION OF BENZOIC ANHYDRIDE

This experiment provides an illustration of a very general and effective method for the preparation of acid anhydrides from acid chlorides. Although this reaction is not often described in elementary texts on organic chemistry, it usually is the method of choice for the preparation of simple anhydrides.

$$C_6H_5-\overset{O}{\overset{\|}{C}}-Cl + C_5H_5N \rightarrow C_5H_5N^{\oplus}-\overset{O}{\overset{\|}{C}}-C_6H_5 \; Cl^-$$

Benzoyl chloride; Pyridine

$$2\, C_5H_5N^{\oplus}-\overset{O}{\overset{\|}{C}}-C_6H_5 \; Cl^- + H_2O \rightarrow \left(C_6H_5-\overset{O}{\overset{\|}{C}}\right)_2 O + 2\, C_5H_5N^{\oplus}-H \; Cl^-$$

Benzoic anhydride

Procedure

PREPARATION OF BENZOIC ANHYDRIDE

Prepare a mixture of 2.0 ml (2.4 g, 0.017 mole) of benzoyl chloride and 8 ml of pyridine (Note 1) in a loosely stoppered 25-ml Erlenmeyer flask. Warm the mixture on the steam bath for 5 minutes and then pour it on 20 g of crushed ice and 10 ml of concentrated hydrochloric acid both in a 100-ml beaker. The anhydride separates at once; however, because of its low melting point it may separate as an oil. Cool the beaker containing the reaction mixture in ice and stir the mixture occasionally with a stirring rod or spatula. As soon as the product solidifies or crystallizes, collect it by suction filtration without allowing the solid to become warm, using a chilled funnel (Note 2). Wash the product on the filter with two 50-ml portions of ice water and continue to pull air through the product for a few minutes to dry. Recrystallize the crude material from petroleum ether or "Skellysolve" (Notes 3, 4). Dry the product, weigh it and determine its melting point. The yield is about 85–90%. Turn in the product with your report.

Note 1. Pyridine has a strong unpleasant smell. It is also somewhat more toxic than many organic solvents. Do not inhale its vapor nor spill the liquid on your skin. If pyridine is accidentally spilled on the skin, wash it off with liberal use of water.

Note 2. If the material on the filter paper appears quite oily or mushy, washing with two 25-ml portions of petroleum ether or "Skellysolve" may be helpful in the final crystallization.

Note 3. One challenging part of this experiment is the recrystallization of the product, m.p. 42–43°. Recrystallization of low-melting solids is more difficult than for higher melting compounds and requires patience, ingenuity, and skill. Often seeding the solution will aid crystallization as will chilling in an ice-salt mixture or in dry ice-ethanol mixture. Sufficient petroleum ether must be added to form a homogeneous solution (one layer only) when the solution is warm.

Note 4. Use caution in the handling of petroleum ether or "Skellysolve." Both have relatively low flash points. Therefore, all operations involving heating these solvents should be done on the steam bath in an area remote from open flames.

TEXT REFERENCE

Report: 23

Chapter Pages

Section Desk NAME

PREPARATION OF *trans*-CINNAMAMIDE

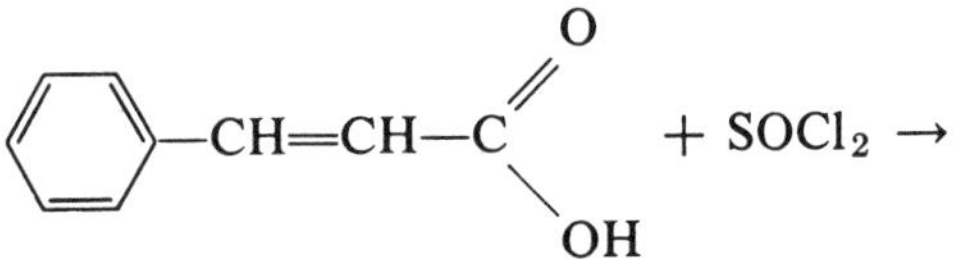

$$C_6H_5\text{—CH=CH—C}(\text{=O})\text{OH} + SOCl_2 \rightarrow$$

Cinnamic acid

$$C_6H_5\text{—CH=CH—C}(\text{=O})\text{Cl} \xrightarrow{NH_4OH} C_6H_5\text{—CH=CH—C}(\text{=O})NH_2$$

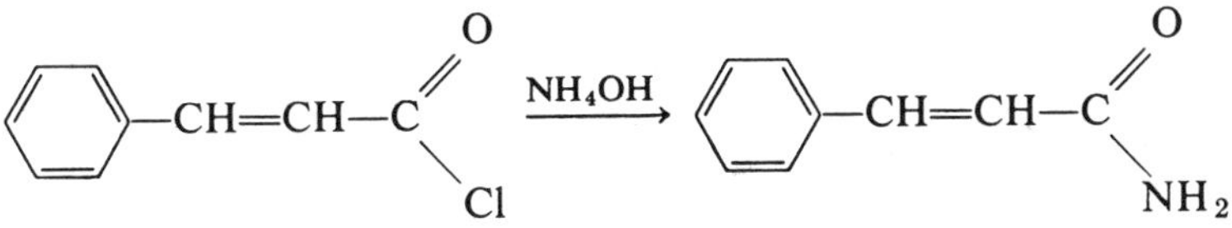

Cinnamoyl chloride Cinnamamide

	Acid	$SOCl_2$	Acyl chloride	Amide
Quantities	3.0 g	2.5 g	______g	______g
Mol. Wt.	______	______	______	______
Moles	______	______	______	______

Theoretical yield of acyl chloride ______g

Actual crude yield of acyl chloride ______g

Theoretical yield of amide ______g

Actual yield of amide ______g

Percentage yield of amide ______

PREPARATION OF PHENYLACETAMIDE

$$C_6H_5\text{—}CH_2\text{—}C(=O)\text{—}OH + SOCl_2 \rightarrow C_6H_5\text{—}CH_2\text{—}C(=O)\text{—}Cl \xrightarrow{NH_4OH} C_6H_5\text{—}CH_2\text{—}C(=O)\text{—}NH_2$$

Phenylacetic acid Phenylacetyl chloride Phenylacetamide

	Acid	$SOCl_2$	Acyl chloride	Amide
Quantities	2.8 g	2.5 g	________g	________g
Mol. Wt.	________	________	________	________
Moles	________	________	________	________

Theoretical yield of acyl chloride ________g

Actual yield of acyl chloride ________g

Theoretical yield of amide ________g

Actual yield of amide ________g

Percentage yield of amide ________

PREPARATION OF BENZOIC ANHYDRIDE

Reaction equation

$$2\,C_6H_5COCl + 2\,C_5H_5N + H_2O \rightarrow (C_6H_5CO)_2O + 2\,C_5H_5NH^{\oplus}\,Cl^-$$

	Benzoyl chloride	Pyridine	Benzoic anhydride
Quantities	2.40 g	8 ml	________g
Mol. Wt.	________		________
Moles	________		________

Theoretical yield ________g

Actual yield ________g

Percentage yield ________

Questions and Exercises

1. Write equations for two different methods of synthesizing *trans*-cinnamamide that do not involve the reaction of an acyl chloride with ammonia.
2. What would be the product of the reaction of phenylacetyl chloride with (a) water, (b) *t*-butylamine, (c) sodium acetate, (d) sodium methoxide, (e) bromine? Write the equations.
3. Write equations showing how phenylacetamide could be converted into (a) phenylacetonitrile, (b) benzylamine.
4. Acyl chlorides have higher molecular weights than the corresponding acids but are much more volatile (phenylacetic acid boils at 257° while phenylacetyl chloride boils at 210°). Why? Why is the difference in volatility less than that between corresponding alcohols and alkyl chlorides?

Experiment 24

Time: 3½–4 hours

Esterification: Preparation of Isoamyl Acetate

Carboxylic acids react with alcohols to form **esters** according to the following equation:

$$R{-}\overset{\overset{\displaystyle O}{\|}}{C}{-}OH + R'{-}OH \rightleftharpoons R{-}\overset{\overset{\displaystyle O}{\|}}{C}{-}O{-}R' + H_2O$$

The esterification reaction involves an equilibrium that is attained only very slowly when the organic acid is allowed to react with the alcohol in the absence of a catalyst. The addition of a strong mineral acid such as sulfuric or hydrochloric acid, while catalyzing the reaction greatly, does not affect the position of the equilibrium. To force the reaction to the right (i.e., to increase the amount of ester formed), two steps can be taken: (1) either reactant (usually the less costly) may be used in excess of the amount called for by the reaction equation, or (2) either of the products may be removed as it is produced. This experiment illustrates the preparation of a typical ester, isoamyl acetate. Isoamyl acetate (bp 142°), sometimes called banana oil, is a relatively common organic solvent used in paints, lacquers, etc.

Procedure

PREPARATION OF ISOAMYL ACETATE

Set up a reflux apparatus like that illustrated in Figure 2.3, page 17, using a 125-ml round-bottomed flask. To the flask add 17.6 g (22 ml, 0.2 mole) of isoamyl alcohol (d = 0.81) and 30 g (30 ml, 0.5 mole) of glacial acetic acid (d = 1.01). Carefully add 4 ml of concentrated sulfuric acid to the mixture. Swirl the flask gently to thoroughly mix the reactants, add a clay chip or "Boileezer" and assemble the apparatus. Start the water flowing through the condenser and heat the mixture to boiling. If heating is done with a burner, adjust the burner to a low flame. Allow the reaction mixture to reflux for one hour. Cool the solution in the reaction flask by immersing it in a cold water bath, then pour it into 100 ml of cold water contained in a 400-ml beaker.

Using a separatory funnel, separate the two layers. (Which is the ester layer?) Return the crude ester to the separatory funnel and wash it with 30 ml of 10% aqueous sodium bicarbonate solution, swirling the mixture in the funnel gently before shaking to permit any carbon dioxide formed to escape. (**CAUTION!** *It is always good practice to make certain that any gas evolution expected is substantially complete before stoppering a separatory funnel and also to release any pressure built up in the funnel at frequent intervals.*) Separate the layers and wash the ester layer by adding 30 ml of water and shaking the mixture for several minutes. Separate the layers and discard the aqueous layer. Transfer the ester to a 50-ml Erlenmeyer flask and dry it for 30 minutes over 3 g of anhydrous magnesium sulfate.

While the ester is drying set up a distillation apparatus using a 100-ml distilling flask (Fig. 3.1). Make certain that all parts of the apparatus are dry (the condenser may still be wet from its use in the reflux apparatus) and see that all connections are tight. Filter the liquid into the distilling flask through a long-stemmed funnel fitted with a small plug of cotton. Add one or two "Boileezers" or pieces of clay plate and distill the ester over a low flame. Collect that fraction boiling from 137–142° in a dry 50-ml flask (Note 1). Weigh your product and submit it to your instructor with your report. Yield, approximately 75%.

Note 1. The vapor pressure of isoamyl acetate at room temperature is sufficiently high that, unless precautions are taken to keep exposure of the product to the air at a minimum, enough vapor will escape into the laboratory to produce an almost heady effect. Therefore, collect your product in a cooled receiver (Fig. 19.1) and lead vapors into the water trough.

TEXT REFERENCE

Report: 24

Chapter Pages

Section Desk NAME

PREPARATION OF ISOAMYL ACETATE

Reaction equation

$$CH_3\overset{O}{\overset{\|}{C}}OH + (CH_3)_2CHCH_2CH_2OH \rightleftharpoons CH_3\overset{O}{\overset{\|}{C}}OCH_2CH_2CH(CH_3)_2 + H_2O$$

Acetic acid — Isoamyl alcohol — Isoamyl acetate

	Acetic acid	Isoamyl alcohol	Isoamyl acetate
Quantities	30.0 g	17.6 g (22 ml)	________
Mol. Wt.	________	________	________
Moles	________	________	________

Theoretical yield ________g

Actual yield ________g

Percentage yield ________

Questions and Exercises

1. Write a mechanism for the acid-catalyzed esterification reaction that will show the role of the sulfuric acid in the preparation of isoamyl acetate.
2. Although the addition of a small amount of sulfuric acid speeds up the attainment of equilibrium in acid-catalyzed esterification, the addition of a large amount of sulfuric acid can result in a slowing down of the attainment of equilibrium and a very low yield of ester. Suggest an explanation for this inhibitory effect of high acid concentration. (*Hint:* A clue may be found in the discussion preceding Experiment 20.)
3. An excellent acid catalyst for many esterifications is anhydrous hydrogen chloride (which is passed into a mixture of the reactants). Why would this catalyst be a poor choice for use with *t*-butyl alcohol? What would be the expected product in such an attempted esterification?
4. The equilibrium constant for many acid-catalyzed esterifications is approximately 4 as is shown by the following expression.

$$K_e = \frac{[\text{products}]}{[\text{reactants}]} = \frac{[\text{ester}][\text{water}]}{[\text{acid}][\text{alcohol}]} \cong 4$$

   Calculate the theoretical yield of isoamyl acetate using the amounts of isoamyl alcohol and acetic acid specified in the experiment and using 4 for K_e.

5. Propose a synthesis for isoamyl acetate that would be a one-way synthesis not involving any equilibrium between reactants and products.

Experiment 25

Time: 2 × 3½ hours

Esterification: Preparation of Aspirin and Oil of Wintergreen

The phenols, unlike the alcohols, cannot be esterified by direct interaction of the phenol with an organic acid. The esterification of a phenol usually is carried out by treating it with an acid anhydride or an acyl chloride. In the first part of this experiment the phenolic acid, salicylic acid, is converted into the acetate ester, aspirin, by treatment with acetic anhydride.

$$\text{C}_6\text{H}_4(\text{OH})\text{–C(=O)–OH} + \text{CH}_3\text{C(=O)–O–C(=O)CH}_3 \xrightarrow{H_3PO_4} \text{C}_6\text{H}_4(\text{O–C(=O)–CH}_3)\text{–C(=O)–OH} + \text{CH}_3\text{C(=O)OH}$$

Salicylic acid | Acetic anhydride | Aspirin

In the second part of the experiment salicylic acid is esterified with methanol at the carboxyl group to yield methyl salicylate (oil of wintergreen). Both aspirin and oil of wintergreen are pharmaceutical chemicals of some importance.

$$\text{C}_6\text{H}_4(\text{OH})\text{–C(=O)–OH} + \text{CH}_3\text{–OH} \xrightarrow{H_2SO_4} \text{C}_6\text{H}_4(\text{OH})\text{–C(=O)–OCH}_3$$

Salicylic acid | Methanol | Methyl salicylate

This experiment also illustrates the use of distillation under reduced pressure for the purification of high-boiling liquids. Distillation of high-boiling liquids at atmospheric pressure is often unsatisfactory because at the high temperatures required the material being distilled may partially (or even totally) decompose resulting in a loss of product and contamination of the distillate. Furthermore, unless the distillation apparatus is properly insulated, thermal losses to the room may be rather large, requiring that the liquid in the still pot be heated even more strongly and slowing down the rate of distillation. However, as was pointed out in the introduction to Experiment 3, when the total pressure (inside the distillation apparatus) is reduced, the boiling point is lowered. Therefore, by varying the pressure under which a distillation is conducted, the temperature required for distillation can be controlled. Distillations carried out in this fashion are called **vacuum distillations** or **distillations under reduced pressure.**

A simple apparatus set-up for distillation under reduced pressure is shown in Figure 25.1. This set-up requires the use of a Claisen adapter. In the first neck of the adapter is inserted a capillary ebullator, the function of which is to admit a *tiny* stream of air (or preferably nitrogen or helium) bubbles to prevent bumping during distillation. "Boileezers" do not function satisfactorily at reduced pressure. The rest of the set-up is fairly typical of simple, atmospheric distillations except that the receiver is fitted with an adapter connected to some pressure reducing device. Two means of reducing pressure are commonly employed in the organic chemistry laboratory, the water aspirator and the oil vacuum pump. The aspirator is useful down to pressures of about 12 mm and the vacuum pump down to pressures below 1 mm. Although the use of an oil pump is no more difficult than the use of an aspirator, it requires more care and rather costly installation; therefore, even in advanced laboratories the aspirator is widely used when the pressure requirement permits.

Some means of measuring the pressure inside the distillation apparatus must be provided. There are many devices available for such measurements, but either a simple manometer or the special manometer shown in Figure 25.1 is suitable for distillations at aspirator pressures. If needed, the apparatus can be modified so that the pressure can be controlled (by "bleeding" in a restricted amount of air at point (c) in the apparatus, for example), but in simple set-ups this feature is usually omitted.

At 760 mm pressure methyl salicylate boils at 222.3°. With care it could be distilled at this pressure, but the process would be more demanding than a distillation under reduced pressure. Figure 25.2 shows the variation in the boiling point of methyl salicylate over the pressure range of about 7 mm to 50 mm. In the range of the water aspirator (12–20 mm) methyl salicylate should distill in the very convenient temperature range of about 100–115°.

All experiments in the organic chemistry laboratory should be carried out with due regard for the hazards involved. When a distillation is conducted under reduced pressure, the apparatus must be treated with more care than in a distillation at atmospheric pressure because of the possibility of an implosion if some part of the glassware in the system being evacuated is broken. Glassware used in such systems should be inspected for flaws or cracks that might rupture under the approximately 15 pounds per square inch pressure applied to the outer surface and essentially zero pressure on the inner surface. Most laboratory glassware designed for the organic chemistry laboratory having cyclindrical or spherical shapes (or nearly so) can be used safely. Many Erlenmeyer flasks are not designed for use under reduced pressure, and any such flask of over

FIGURE 25.1 Vacuum distillation assembly. A short section of rubber tubing should be stretched over the top of the capillary ebullator. A short length of copper wire inserted into the tubing will prevent a complete cut off of air when clamp is closed.

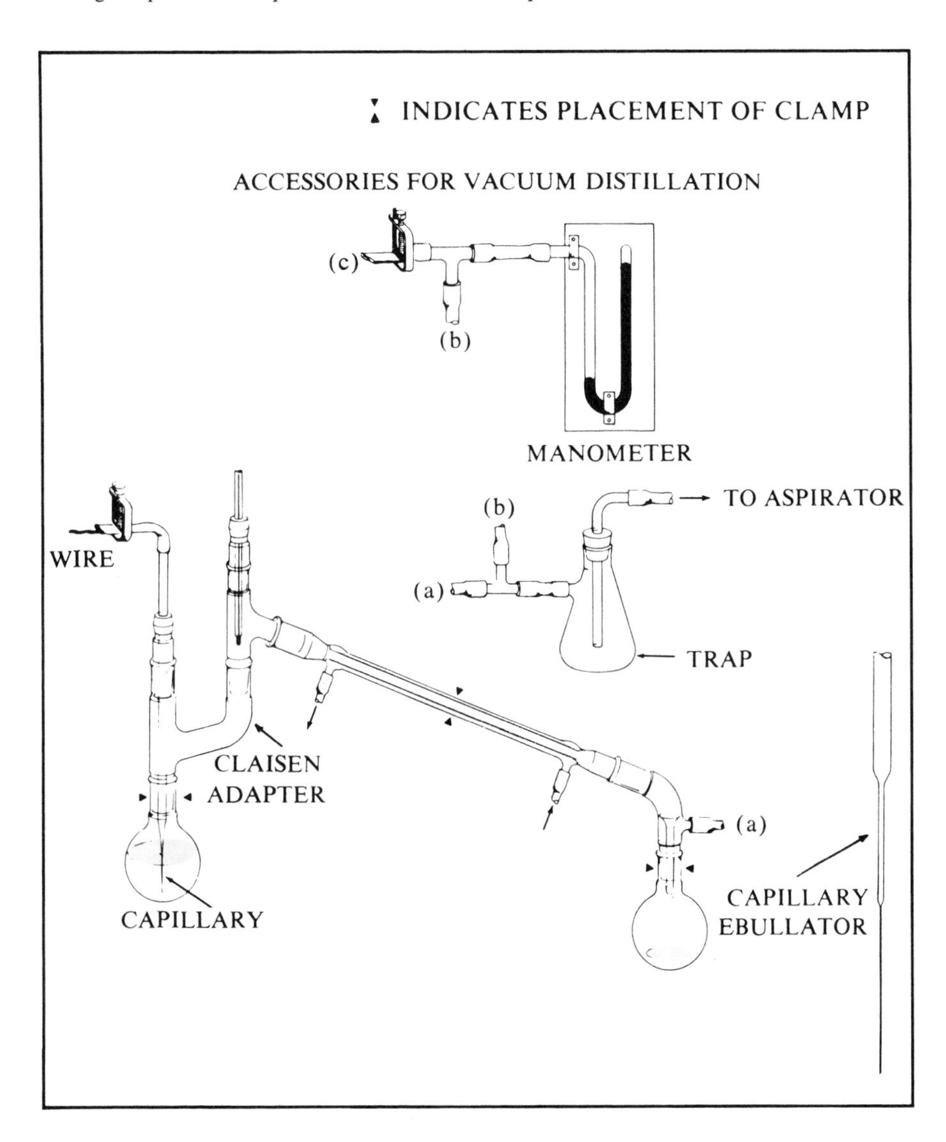

50-ml capacity should not be evacuated unless it is known that the flask was intended for use under reduced pressure. If adjustments are required in a system being evacuated, the pressure should be allowed to return to its normal value

FIGURE 25.2 The boiling point of methyl salicylate at various pressures.

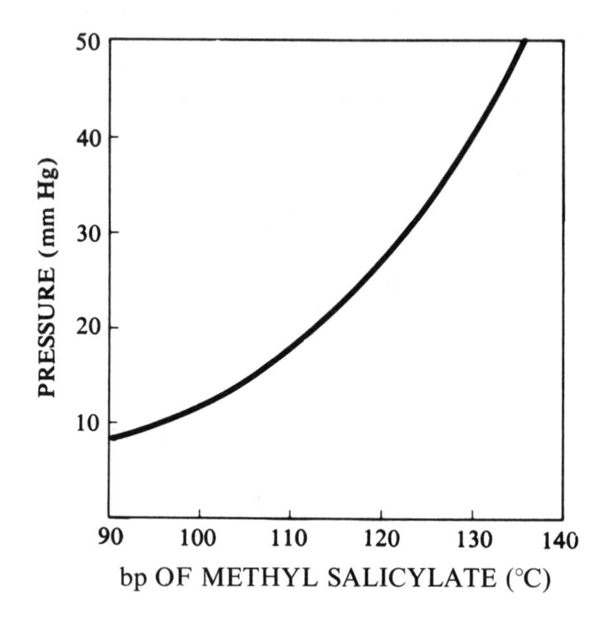

before moving, twisting, or otherwise applying force to the system (except for adjustable components built into the system such as stop-cocks, fraction collectors, etc.). Safety glasses must be worn while carrying out a vacuum distillation.

PART I PREPARATION OF ASPIRIN

Procedure

If you are assigned both parts of this experiment, you can do them simultaneously by starting Part II first and then carrying our Part I while the reaction mixture in Part II is being heated under reflux 3–3½ hours.

A. PREPARATION OF SALICYLIC ACID ACETATE

Place 2 g of salicylic acid in a 125-ml Erlenmeyer flask, add 5 ml of acetic anhydride and 5 drops of 85% phosphoric acid. Stir the mixture well. Heat the flask in the boiling water bath for 5 minutes, remove from the bath and, while still hot, *cautiously* add 2 ml of water in one portion. **(CAUTION!** ***The*** *solution may boil from the heat of decomposition of the excess acetic anhydride;* **handle the flask carefully.)** After decomposition is complete, add 40 ml of water and stir the solution until crystals begin to form. Cool the mixture in the ice bath to complete the crystallization. Collect the product by suction filtration on the Hirsch funnel, wash with 5 ml of cold water and pull air through the filter until the product is dry. Recrystallize the product from 35 ml of hot water (Note 1), using decolorizing charcoal if the product is colored. Dry the product and determine its melting point. Perform the test described below and turn in your product together with your report.

B. TEST FOR THE PHENOLIC HYDROXYL GROUP

Dissolve a few crystals of aspirin in 1 ml of methanol in a test tube and add one drop of 1% ferric chloride solution. Record your result. Repeat the test with salicylic acid. Result?

Note 1. The water should not be heated above 80°. Boiling water will partially hydrolyze aspirin to salicylic acid and acetic acid.

PART II PREPARATION OF METHYL SALICYLATE

Procedure

A. ESTERIFICATION OF SALICYLIC ACID

Set up a reflux apparatus like that illustrated in Figure 2.3, using a 100-ml round-bottomed flask. To the flask add 6.9 g (0.05 mole) of salicylic acid and 24 g (30 ml, 0.75 mole) of methyl alcohol. *Carefully* add in small portions 8 ml of concentrated sulfuric acid to the mixture. Swirl the flask gently after each addition of acid to thoroughly mix the reactants. Add a clay chip or "Boileezer" to the flask, and assemble the apparatus. Start the water flowing through the condenser and heat the mixture to boiling over a low flame. Allow the mixture to reflux for 3–3½ hours (Note 1). Cool the solution in the reaction flask by immersing it in a cold water bath, then add 50 ml of water. Pour the mixture into a 125-ml separatory funnel. Rinse the reaction flask with 25 ml of ether and add the ether rinsings also to the separatory funnel. Shake well and separate the layers. Extract the aqueous layer with a second 25-ml portion of ether and combine the ether layers. Discard the aqueous layer. Wash the ether solution once with 25 ml of a 5% $NaHCO_3$ solution and once with 25 ml of a saturated sodium chloride solution. Separate and store the ether solution in a 125-ml Erlenmeyer flask over 1–2 g of anhydrous calcium chloride until the next laboratory period.

B. DISTILLATION UNDER REDUCED PRESSURE[1]

CAUTION! *See last paragraph of discussion on distillation under reduced pressure, pages 224 and 226.*

Set up an apparatus for distillation under reduced pressure by assembling the equipment illustrated in Figure 25.1 using a 50-ml round-bottomed flask fitted with a Claisen adapter. Fit the distilling flask with a capillary ebullator tube reaching to the bottom. A capillary ebullator tube may be made by heating a length of 6-mm glass tubing in your burner until it is very soft, removing it from the flame, and rapidly drawing it out to a diameter slightly smaller than that of a capillary melting point tube and an added length of 8 to 12 inches. Heat the capillary at a point about 2 inches from its end in a *low* flame and rapidly draw it out to a diameter about that of a hair. Break the fine capillary to a length of 2 inches or longer (Note 2). If a bleeder tube is supplied with the ground-glass equipment kit, it will serve as a capillary if equipped as illustrated in Figure 25.1. Connect the side arm of the take-off adapter to the glass "tee" [labelled (a) in Fig. 25.1] leading to the manometer and trap and connect the trap to the water aspirator.

Introduce the dried ether solution of methyl salicylate into the distilling flask via a long-stemmed funnel. Make certain that all connections and screw clamps are tight. Place the apparatus under diminished pressure by turning on the water aspirator full force and remove the ether, using a warm (40°) water bath as the heat source. (**CAUTION!** *No open flames!*) A thin stream of bubbles should come out of the ebullator tube. When all the ether has been removed (no more distillate appears to pass over), stop heating, open the screwclamp or stopcock at the manometer "tee" [labelled (c) in Figure 25.1], and turn off the water. Always "break the vacuum" before turning off the aspirator. Change receivers and resume the distillation under diminished pressure, distilling the ester over a low flame. Collect all distillate that passes over at a temperature of 100° or above (Note 3). When no more distillate passes over, stop heating and open the apparatus to the atmosphere as described previously. Weigh the product in a dry 25-ml Erlenmeyer flask, label, and submit it to your instructor with your report. Yield of pure product will be approximately 5.0 grams (66%).

[1]The water pressure in some laboratories is greatly diminished when all students attempt to use aspirators at the same time. If vacuum pumps are available, some instructors may prefer to modify the procedure under Part B to adapt it to the use of a high vacuum device. In this event, the trap should be one cooled by "dry ice" and the distillation carried out using the combined ester residues of 5–6 students.

Note 1. A long reflux time is required to esterify salicylic acid and obtain a respectable yield. A supplementary experiment may be performed during this reaction period, or work from previous experiments still pending may be completed.

Note 2. The capillary ebullator may be tested by blowing through it into a small quantity of acetone in a beaker. If the capillary is of the proper size a thin stream of fine bubbles will be seen. If no bubbles form the capillary is too small. If a coarse stream of bubbles is seen, the capillary is too large.

Note 3. The normal boiling point of methyl salicylate is 224° but the ester will distill at 105° at a pressure of 14 mm. A pressure of 14 mm is about the limit obtainable with a good aspirator and a water temperature of 16.5°. If a different pressure is obtained, the corresponding temperature at which the ester will boil may be determined from the curve, Figure 25.2.

TEXT REFERENCE

Report: 25

Chapter Pages

Section Desk NAME

PREPARATION OF ASPIRIN (SALICYLIC ACID ACETATE)

$$C_6H_4(OH)COOH + (CH_3CO)_2O \xrightarrow{H_3PO_4} C_6H_4(OCOCH_3)COOH + CH_3COOH$$

Salicylic acid + Acetic anhydride → Aspirin

	Salicylic acid	Acetic anhydride	Aspirin
Quantities	2.0 g	(5 ml) 5.41 g	______g
Mol. Wt.	______	______	______
Moles	______	______	______

Theoretical yield ______g

Actual yield ______g

Percentage yield ______

Result of the ferric chloride test on aspirin ______

Result of the ferric chloride test on salicylic acid ______

PREPARATION OF METHYL SALICYLATE (OIL OF WINTERGREEN)

$$C_6H_4(OH)COOH + CH_3OH \xrightarrow{H_2SO_4} C_6H_4(OH)COOCH_3$$

	Salicylic acid	Methanol	Methyl salicylate
Quantities	6.9 g	24 g	______ g
Mol. Wt	______	______	______
Moles	______	______	______

Theoretical yield ______ g

Actual yield ______ g

Percentage yield ______

Partners in experiment:

(1) ______

(2) ______

(3) ______

Questions and Exercises

1. A third pharmaceutical product derived from salicylic acid is salol (phenyl salicylate), the phenyl ester of salicylic acid. Since phenol cannot be esterified by direct interaction with salicylic acid, some indirect method must be used. Write equations for a sequence that might be used to prepare salol from phenol and salicylic acid.
2. Many of the more expensive vacuum distillation set-ups avoid the use of a capillary ebullator, substituting a magnetic stirrer or a mechanical stirrer (driven by a magnet). What is one disadvantage of the capillary ebullator that could make it difficult to obtain a pure product by the vacuum distillation of certain compounds? How can we modify the ebullator shown in Figure 25.1 to avoid this difficulty?
3. Aspirin and salol are both acidic substances. Which is the stronger acid? Which would be more easily hydrolyzed in an alkaline medium (like that in the intestine)? Why?
4. For reasons of expedience, the distillation of methyl salicylate under reduced pressure was carried out using a low flame as a heat source. Actually, a more accurate relationship between pressure and boiling point is achieved when heating is done with an electric mantle or an oil bath. Why?
5. What would be the pressure if methyl salicylate were brought to a boil at 140°C?

Experiment 26

Time: 1½ hours

The Preparation of Acetanilide

Ammonia reacts with acyl halides, anhydrides, or esters to yield the corresponding **amides.**

$$R-\overset{\overset{O}{\|}}{C}-Cl + NH_3 \rightarrow R-\overset{\overset{O}{\|}}{C}-NH_2 + HCl$$

$$(R-\overset{\overset{O}{\|}}{C})_2O + NH_3 \rightarrow R-\overset{\overset{O}{\|}}{C}-NH_2 + R-\overset{\overset{O}{\|}}{C}-OH$$

$$R-\overset{\overset{O}{\|}}{C}-OR' + NH_3 \rightarrow R-\overset{\overset{O}{\|}}{C}-NH_2 + R'-OH$$

If one or more of the hydrogen atoms of the ammonia molecule is replaced by an alkyl group, the resultant compound is an **amine.** Amines that have at least one hydrogen atom on the nitrogen atom, like ammonia, react with the above acid derivatives to yield **N-substituted amides.** N-substituted amides in which the hydrocarbon group on the nitrogen atom is a phenyl group, C_6H_5—, are called **anilides.** In the present experiment aniline is treated with acetic anhydride to produce acetanilide.

$$\underset{\text{Aniline}}{C_6H_5-NH_2} + (CH_3-\overset{\overset{O}{\|}}{C})_2O \rightarrow \underset{\text{Acetanilide}}{C_6H_5-\overset{\overset{H}{|}}{N}-\overset{\overset{O}{\|}}{C}-CH_3} + CH_3-\overset{\overset{O}{\|}}{C}-OH$$

Procedure

Dissolve 20 ml of concentrated hydrochloric acid in 250 ml of water and add 20 g (20 ml, 0.222 mole) of aniline, stirring until the aniline dissolves (a beaker may be used for this step). If the aniline is badly discolored, add 2 g of decolorizing charcoal, stir the mixture for a few minutes, and filter with suction through the Büchner funnel. Dissolve 32 g of sodium acetate crystals ($CH_3CO_2Na \cdot 3\ H_2O$) in 100 ml of water. Filter the sodium acetate solution if necessary to remove undissolved particles.

Pour the solution of aniline hydrochloride into a 600-ml beaker, support on a wire gauze and gently warm the solution over a low flame while stirring it with a thermometer. When the temperature of the solution reaches 50°, remove the beaker from the ring, place it on the desk, and add 24 ml of acetic anhydride. (**CAUTION!** *The vapors of acetic anhydride are very irritating to the nose, throat, and eyes.*) Stir the solution until the acetic anhydride dissolves and *immediately* add in one portion the solution of sodium acetate. Stir the mixture vigorously. Crystals of pure acetanilide should begin to precipitate almost immediately. Cool the reaction mixture in an ice bath and continue to stir the mixture while the product crystallizes. Collect the crystals by filtering them with suction on the Büchner funnel, wash them with 20 ml of *cold* water, and press the crystal cake on the filter with a clean rubber stopper. Allow the crystals to dry until the next laboratory period. The acetanilide you have prepared may be used in following experiments without further purification. Protect the crystals from dust, dirt, etc. with a piece or clean filter paper or paper towel.

Weigh the dried product and determine its melting point. Transfer your product into a clean, dry, labeled bottle and turn it in with your written report. Your instructor will return your preparation for use in Experiment 32. Yield, 14–16 g; mp 115°.

TEXT REFERENCE

Report: 26

Chapter Pages

Section Desk NAME

PREPARATION OF ACETANILIDE

Reaction equation

$$C_6H_5\text{—}NH_2 + (CH_3\text{—}\overset{O}{\overset{\|}{C}})_2O \rightarrow C_6H_5\text{—}\overset{H}{\overset{|}{N}}\text{—}\overset{O}{\overset{\|}{C}}\text{—}CH_3 + CH_3\text{—}\overset{O}{\overset{\|}{C}}\text{—}OH$$

	Aniline	Acetic anhydride	Acetanilide
Quantities	20 ml (20 g)	24 ml (26 g)	________g
Mol. Wt.	________	________	________
Moles	________	________	________

Theoretical yield ________g

Actual yield ________g

Percentage yield ________

mp ________°C

Questions and Exercises

1. Although ethyl phenylacetate reacts readily with ammonia and with methylamine to yield the corresponding amide, the reaction of ethyl phenylacetate with *t*-butylamine is so slow and the yield so small as to render the reaction useless in the laboratory. Suggest a reason for the failure of the aminolysis with *t*-butylamine. (*Hint:* What is the mechanism of aminolysis?)
2. Write equations for the syntheses of the following amides: (a) benzanilide, (b) N-*t*-butylphenylacetamide, (c) succinimide.
3. What is the function of the sodium acetate used in the preparation of acetanilide according to the procedure given? Write the equation for the preparation of acetanilide as you have actually performed the experiment. (Not the equation on the report form!) Include all reactants and indicate the role of sodium acetate.
4. Acetyl chloride and ketene, $H_2C{=}C{=}O$, may be used to acetylate aniline but both reagents offer certain disadvantages. Suggest possible disadvantages to the use of each. Do these reagents offer any obvious advantages?
5. For persons sensitive to aspirin, 4-hydroxyacetanilide often is prescribed as an analgesic. Suggest a synthesis leading to the preparation of this pharmaceutical product beginning with phenol.

Experiment 27

Time: 3 hours

The Friedel-Crafts Reaction: Preparation of p-Anisyl Benzyl Ketone or p-Methoxybenzophenone

Although the Friedel-Crafts reaction is a very general reaction in which an unsaturated hydrocarbon (alkene, alkyne, or aromatic hydrocarbon), or certain simple derivatives thereof, is alkylated or acylated using a Lewis acid catalyst, the most commonly encountered examples of this method involve the alkylation or acylation of aromatic hydrocarbons, aryl halides, or aryl ethers. A variety of alkylation agents may be used, alkyl halides, alcohols, and alkenes being the most common. Acyl halides and acid anhydrides are the most frequently employed acylation agents, but, as is illustrated in this experiment, under the proper conditions carboxylic acids may also be used. Aluminum chloride is probably the most common "Friedel-Crafts catalyst" and, for that reason, appears in most general equations for the Friedel-Crafts reaction. Other catalysts include zinc chloride, stannic chloride, sulfuric acid, hydrofluoric acid, etc. In practice the choices of aromatic substrate, alkylating or acylating agent, and catalyst are all interdependent, and the selection of one component in the reaction may limit possible choices for the remaining components.

This experiment illustrates the acylation of an aromatic ether (anisole) using a carboxylic acid (phenylacetic acid or benzoic acid) as the acylating agent and polyphosphoric acid (PPA) as the catalyst.

$$C_6H_5-CH_2-\overset{O}{\overset{\|}{C}}-OH + C_6H_5-OCH_3 \xrightarrow{PPA} C_6H_5-CH_2-\overset{O}{\overset{\|}{C}}-C_6H_4-OCH_3$$

Phenylacetic acid Anisole

$$C_6H_5-\overset{O}{\overset{\|}{C}}-OH + C_6H_5-OCH_3 \xrightarrow{PPA} C_6H_5-\overset{O}{\overset{\|}{C}}-C_6H_4-OCH_3$$

Benzoic acid

The use of polyphosphoric acid as a Friedel-Crafts catalyst is a relatively recent innovation. In many reactions it has the advantage of serving as an effective acid catalyst, serving as solvent for the reaction, resulting in high yields with few by-products, and simplifying the equipment set-up and the work-up of product. In the more conventional Friedel-Crafts acylation shown in the following equation either an excess of the aromatic substrate or some inert

$$R{-}\overset{\overset{\large O}{\|}}{C}{-}X + C_6H_5{-}R' \xrightarrow{AlCl_3} R{-}\overset{\overset{\large O}{\|}}{C}{-}C_6H_4{-}R' + HX$$

$$X = Cl \text{ or } R{-}\overset{\overset{\large O}{\|}}{C}{-}O$$

organic compound must be used as solvent. The reaction is often heterogeneous, and provision must be made for the handling of anhydrous aluminum chloride and the trapping of escaping hydrogen halides (from the aluminum chloride and/or the acylating agent). There is no real difficulty in carrying out reactions of this type; however, the simplicity and ease of carrying out the polyphosphoric acid-catalyzed Friedel-Crafts reaction recommends its use whenever it is applicable.

Note that, whatever the catalyst, the predominant product in acylations of this type is the *para*-isomer. Generally, very little or none of the *ortho*- or *meta*-isomers is obtained.

Your instructor will assign you either procedure (A) or (B) of this experiment. The method used in procedure (A) is simpler, but it requires that you work with phenylacetic acid (Note 1), an important and relatively safe carboxylic acid but, unfortunately, one that shares with its low molecular weight aliphatic relatives (valeric through capric acid) a very unpleasant odor. The ability to work with phenylacetic acid without carrying its characteristic odor away from the laboratory is said by some to be indicative of superior laboratory technique.

Procedure

CAUTION! *Polyphosphoric acid is a very viscous liquid, which is not easy to pour, particularly when cold. When in use, it should be kept in a reasonably warm place and should be handled with care. If any of the acid is spilled on the bench or comes into contact with any part of the body or clothing, it should be removed at once by washing the affected area with a large volume of water, followed by dilute aqueous sodium bicarbonate, and more water. Phenylacetic acid presents no particular hazard, but it has a persistent, unpleasant odor; therefore, care should be taken to avoid getting it on the skin or clothing* (*Note* 1).

A. PREPARATION OF *p*-ANISYL BENZYL KETONE

In a 125-ml flask place 15 g of polyphosphoric acid, 1.4 g of phenylacetic acid (Note 1), and 1.2 g of anisole. Loosely stopper the flask with a clean rubber stopper and place it in a boiling-water bath (Note 2). After a minute or two remove the flask and swirl it vigorously to mix the contents and obtain a homogeneous reaction mixture. Return the flask to the bath and allow it to remain there for 45 minutes, except for brief periods at about 10-minute intervals during which the flask is removed from the bath, swirled vigorously, and returned to the bath. The reaction mixture should change color from colorless to deep red as the reaction progresses.

After the heating period is complete, take the flask from the hot water bath, remove the rubber stopper, and chill the flask in an ice bath. Add 50 ml of cold water, stopper tightly and shake the flask vigorously to hydrolyze the polyphorphoric acid-product complex and any excess polyphosphoric acid. Disappearance of the red color may be used as an indicator of the completeness of the hydrolysis of the complex. Chill the flask thoroughly to complete precipitation of the product. Collect the crude product by suction filtration on the Hirsch funnel; wash with 5 ml of cold water, 5 ml of 10% aqueous sodium bicarbonate, and a second 5 ml of cold water; and draw air through the filter until the product is reasonably dry. Recrystallize the product from methanol (using decolorizing charcoal if the product is colored). Dry the product and determine its melting point. Approximate yield, 1.5 to 2 g, mp 75°.

Note 1. See Note 2 of Experiment 23-A.

Note 2. A metal boiling-water bath is preferable from the standpoint of safety. If a metal bath is not available, a 400-ml beaker half filled with water and heated by a burner or, preferably, an electric hot plate will suffice.

B. PREPARATION OF *p*-METHOXYBENZOPHENONE
(See preceding CAUTION on use of PPA)

In a 125-ml flask place 20 g of polyphosphoric acid, 1.6 g of anisole, and 2.4 g of benzoic acid. Stopper the flask *loosely* with a rubber stopper and support it with a clamp in a boiling-water bath. After a minute or two, remove the flask and swirl to thoroughly mix the reactants. Return the flask to the water bath and continue heating for a period of one hour, removing the flask every ten minutes or so to remix the contents by swirling. The reaction mixture should change color from colorless to amber.

After the heating period is complete, take the flask from the water bath, remove the rubber stopper, and chill the flask in an ice bath. Add 50 ml of cold water, stopper loosely, and swirl the flask vigorously to hydrolyze the polyphosphoric acid-product complex and any excess polyphosphoric acid. Disappearance of the amber color may be used as an indicator of the completeness of the hydrolysis of the complex. Chill the flask thoroughly to complete precipitation of the product. Extract the crude product, which has the form of a waxy solid, by adding to the flask 25 ml of ether. The solution of solid material by the ether may be hastened by warming in a water bath and swirling. Transfer the entire contents of the flask to a separatory funnel and separate the ethereal layer. Discard the lower aqueous layer. Wash the ether layer with a 50-ml portion of 10% aqueous sodium hydroxide solution and then with 50 ml of water. Discard both the basic and water washes. Transfer the ether layer to an evaporating dish and evaporate the ether from a steam bath. (**CAUTION!** *See* Note 1.) The syrupy liquid residue will solidify on cooling. Yield about 2.5 g. The crude ketone may be recrystallized from methanol, mp 62°.

Note 1. The evaporation of ether must be carried out in the exhaust hood. All flames in the vicinity must be extinguished. *Do not evaporate ether from a hot plate.*

TEXT REFERENCE

Report: 27

Chapter Pages

Section Desk NAME

FRIEDEL-CRAFTS REACTION: PREPARATION OF *p*-ANISYL BENZYL KETONE

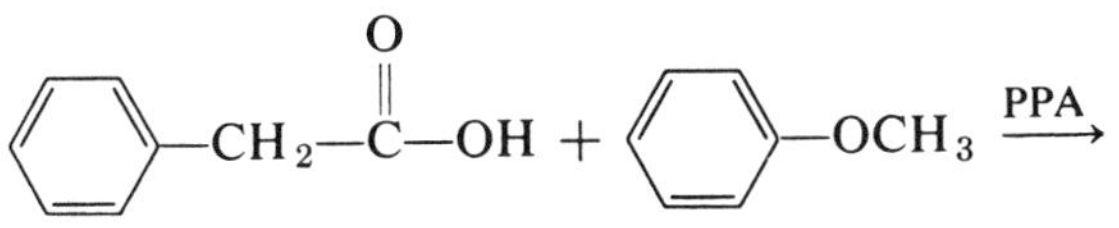

$$C_6H_5-CH_2-\overset{\displaystyle O}{\overset{\|}{C}}-OH + C_6H_5-OCH_3 \xrightarrow{PPA}$$

Phenylacetic acid Anisole

$$C_6H_5-CH_2-\overset{\displaystyle O}{\overset{\|}{C}}-C_6H_4-OCH_3$$

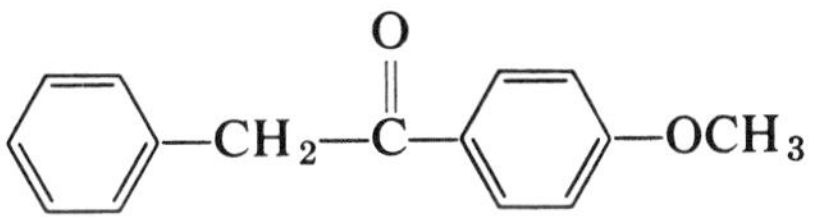

p-Anisyl benzyl ketone

	Phenylacetic acid	Anisole	*p*-Anisyl benzyl ketone
Quantities	1.4 g	1.2 g	________g
Mol. Wt.	________	________	________
Moles	________	________	________

Theoretical yield ________g

Actual yield ________g

Percentage yield ________

PREPARATION OF *p*-METHOXYBENZOPHENONE

$$C_6H_5-\overset{O}{\overset{\|}{C}}-OH + C_6H_5-OCH_3 \xrightarrow{PPA} C_6H_5-\overset{O}{\overset{\|}{C}}-C_6H_4-OCH_3$$

Benzoic acid — Anisole — *p*-Methoxybenzophenone

	Benzoic acid	Anisole	*p*-Methoxybenzophenone
Quantities	2.4 g	1.6 g	______ g
Mol. Wt.	______	______	______
Moles	______	______	______

Theoretical yield ______ g

Actual yield ______ g

Percentage yield ______

Questions and Exercises

1. What is the function of the Lewis acid catalyst in the Friedel-Crafts acylation reaction?
2. In the Friedel-Crafts reaction only catalytic amounts of catalyst are required for alkylation; however, a full molar equivalent or more of catalyst is required for acylation. Suggest an explanation.
3. On the basis of your answers to the first two questions, suggest an explanation for the appearance of the red color during the course of your experiment.
4. Why does the *p*-isomer predominate in the product of Friedel-Crafts acylations?
5. If *p*-anisyl benzyl ketone were treated with a nitrating mixture, what product would most likely result?
6. What are some of the limitations of the Friedel-Crafts reaction?
7. Why is nitrobenzene sometimes used as a solvent for the Friedel-Crafts reaction?

Experiment 28

Time: 2 hours

A Molecular Rearrangement: The Preparation of Benzanilide from Benzophenone

An important class of molecular rearrangements is that in which an alkyl or aryl group with its pair of bonding electrons migrates to an electron-deficient carbon, oxygen, or nitrogen atom. Such rearrangements to nitrogen are usually irreversible and often are stereospecific.

The rearrangement of ketoximes when catalyzed by acids is a classic example of a molecular rearrangement. It is called the **Beckmann rearrangement** after the German chemist, who in 1886 was the first to discover that the oximes of certain carbonyl compounds when treated with acids rearranged to amides.

$$(Ar)(Ar')C{=}O + H_2NOH \cdot HCl \rightarrow (Ar)(Ar')C{=}N{-}OH + H_2O$$

$$(Ar)(Ar')C{=}N{-}OH + H^+ \rightarrow (Ar)(Ar')C{=}N{-}\overset{+}{O}H_2 \rightarrow Ar{-}C{\equiv}\overset{+}{N}{-}Ar' + H_2O$$

$$Ar{-}C{\equiv}\overset{+}{N}{-}Ar' + H_2O \rightarrow Ar{-}\underset{}{\overset{OH}{\overset{|}{C}}}{=}N{-}Ar' \rightleftharpoons Ar{-}\overset{O}{\overset{\|}{C}}{-}\overset{H}{\overset{|}{N}}{-}Ar' + H^+$$

The mechanism usually proposed for the Beckmann rearrangement is that of a concerted reaction in which the migration of one group is accompanied by the departure of another.

An inspection of the reaction equation above also will show that the carbon-nitrogen double bond has the necessary geometry to permit the formation of two different stable oximes from an unsymmetrical ketone. Such isomers are called *syn* and *anti* and are related respectively to *cis* and *trans* forms of isomers first encountered in doubly-bonded carbon-carbon structures. Regardless of the oxime isomer formed, the migrating group is always the one that is *anti* to the hydroxyl group. The orientation of the groups in the starting oxime, therefore, may be determined from an examination of the resultant amide.

Beckmann used phosphorus pentachloride to effect the rearrangement of the oximes in his original investigation. However, the rearrangement of oximes may be accomplished by any one of a number of acids. In the present experiment we will illustrate the facility with which the rearrangement may be accomplished with polyphosphoric acid.

Procedure

CAUTION! *Polyphosphoric acid is a very viscous liquid, which is not easy to pour, particularly when cold. When in use, it should be kept in a reasonably warm place and should be handled with care. If any of the acid is spilled on the bench or comes into contact with any part of the body or clothing, it should be removed at once by washing the affected area with a large volume of water, followed by dilute aqueous sodium bicarbonate, and more water.*

In a 125-ml Erlenmeyer flask place 20 g of polyphosphoric acid, 1.82 g of benzophenone, and 2.1 g of hydroxylamine hydrochloride. Place the flask in a boiling-water bath and for a minute or two rotate and swirl in order to thoroughly mix the reactants. Clamp the flask in position and continue heating for 30–45 minutes or until frothing has ceased (Note 1). At the end of the reaction period add 50–75 g of crushed ice to the reaction mixture and shake. Benzanilide will begin to precipitate almost immediately. After all the ice has melted, collect your product on the Büchner funnel. Wash twice with 10-ml portions of ice water, dry, and weigh. The yield is 1.5–1.7 g; mp 161°. The benzanilide will be reasonably pure as collected but may be recrystallized from hot ethanol.

Note 1. Hydrogen chloride is released as the reaction takes place. For this reason the reaction should be carried out under an exhaust hood to prevent fumes from being released into the laboratory.

TEXT REFERENCE

Report: 28

Chapter Pages

Section Desk NAME

THE PREPARATION OF BENZANILIDE FROM BENZOPHENONE

Reaction equation

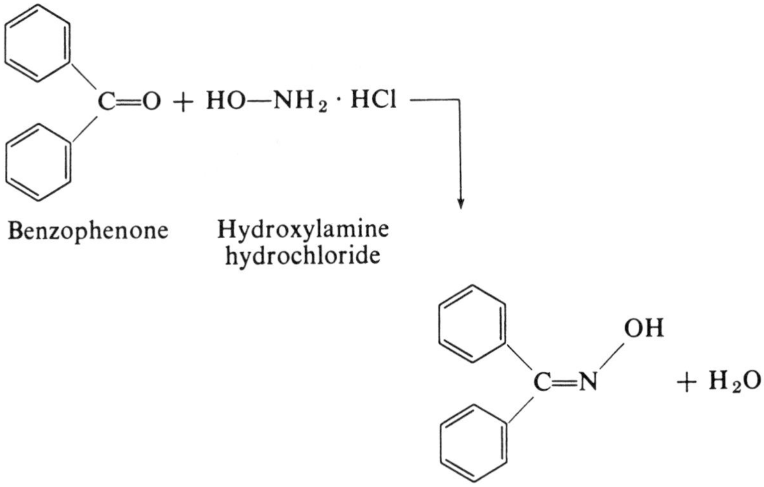

$$(C_6H_5)_2C{=}O + HO{-}NH_2 \cdot HCl \longrightarrow (C_6H_5)_2C{=}N{-}OH + H_2O$$

Benzophenone Hydroxylamine hydrochloride

Benzophenone oxime

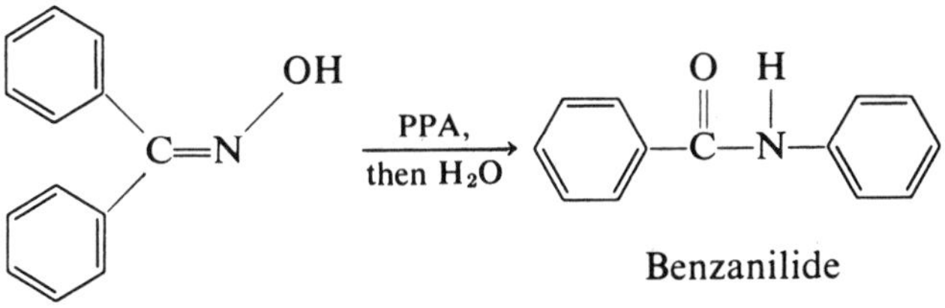

$$(C_6H_5)_2C{=}N{-}OH \xrightarrow[\text{then } H_2O]{\text{PPA,}} C_6H_5{-}\overset{O}{\overset{\|}{C}}{-}\overset{H}{\overset{|}{N}}{-}C_6H_5$$

Benzanilide

Quantities	1.82 g	______
Mol. Wt.	______	______
Moles	______	______

Theoretical yield ______ g

Actual yield ______ g

Percentage yield ______

mp ______ °C

Questions and Exercises

1. Outline a procedure for the preparation and isolation of aniline from benzophenone.
2. What products would be possible from a Beckmann rearrangement of acetophenone?
3. Aliphatic aldehydes and ketones do not give isomeric oximes. Suggest a reason for this.
4. If the oxime of cyclohexanone were treated with acid, what would be the structure and name of the rearranged product?
5. Nylon 6-6, $-\overset{H}{\overset{|}{N}}\!\left[\overset{O}{\overset{\|}{C}}-(CH_2)_4-\overset{O}{\overset{\|}{C}}-\overset{H}{\overset{|}{N}}-(CH_2)_6-\overset{H}{\overset{|}{N}}\right]_n$, may be manufactured from cyclohexanol. One of the necessary steps in the process is a Beckmann rearrangement. Write a series of reactions that would lead to the production of this linear polymer.

Experiment 29

Time: 2 hours

Polymerization

Polymerization is a reaction in which many single molecules (monomers) are united to form giant molecules or **polymers.** The latter are high molecular weight compounds which often possess properties that make them very useful as structural materials. Reactions leading to the production of polymers are of two principal types. If the polymer is formed by a chain reaction of the monomer, the reaction is referred to as **chain** (or **addition**) **polymerization.** Common examples of polymers formed in this manner are polyethylene $\left(CH_2CH_2 \right)_n$, "Teflon", $\left(CF_2CF_2 \right)_n$, and many of the very useful vinyl polymers, $\left(CH_2-\underset{X}{\overset{H}{C}} \right)_n$. Chain polymerization often proceeds via a *free-radical* chain reaction mechanism catalyzed by a peroxide, ultraviolet light, or some agent which aids in the formation of a free radical. The reaction involves three different steps: (1) initiation, (2) propagation, and (3) termination.

Initiation:

$$CH_2{=}CH_2 + R\cdot \rightarrow R-CH_2-CH_2\cdot$$

Propagation:

$$RCH_2-CH_2\cdot + n\,CH_2{=}CH_2 \rightarrow R\left(CH_2-CH_2 \right)_n CH_2CH_2\cdot$$

Termination:

$$R\left(CH_2-CH_2 \right)_n CH_2CH_2\cdot + R\cdot \rightarrow R\left(CH_2CH_2 \right)_{n+1}R$$

When the amount of initiator (R, the free radical) is small and n is a large number (1000–2000), the molecular weight of the polymer becomes very great. Another type of polymerization reaction leading to the formation of high molecular weight compounds is **step** (or **condensation**) **polymerization.** Interaction of functional groups in this type of polymerization reaction takes place in a stepwise fashion, usually producing small molecules (water, hydrogen chloride, ammonia) as byproducts in each step. Cellulose and proteins are examples of natural polymers of this type. Restoring the simple molecules (water, in this case) by hydrolysis of these natural substances results in the formation of the unit molecules from which they were formed (i.e., simple sugars and α-amino acids, respectively). Examples of step polymers familiar to almost everyone are the synthetic fibers "Nylon", a polyamide; and "Dacron", a polyester.

There are a number of ways in which a step polymer such as Nylon can be prepared; however, the procedure that is most suitable for the beginning laboratory student is based on the reaction of a primary amine with an acyl chloride to yield an amide. Thus, the reaction of a diacid chloride with a diamine can proceed under the proper conditions to form a polyamide in which the amide linkage, $-\overset{O}{\overset{\|}{C}}-\overset{H}{\overset{|}{N}}-$, recurs repeatedly to produce a linear polymer of approximately 12,000 molecular weight. Nylon 6-6 and Nylon 6-10 are examples of such linear polyamides, in which the 6-6 and 6-10 designations refer to the 6-carbon diamine and the 6- and 10-carbon diacids, respectively, from which the polymers originate. In procedure (B) of the present experiment Nylon 6-10 is prepared from sebacoyl chloride and hexamethylenediamine by a procedure that is sometimes called "The Nylon Rope Trick."

$$n\ \mathrm{ClC(=O)-(CH_2)_8-C(=O)Cl} + n\ \mathrm{H_2N-(CH_2)_6-NH_2} \rightarrow$$

Sebacoyl chloride Hexamethylenediamine

$$\mathrm{Cl}\left[\overset{O}{\overset{\|}{C}}-(CH_2)_8-\overset{O}{\overset{\|}{C}}-\overset{H}{\overset{|}{N}}-(CH_2)_6-\overset{H}{\overset{|}{N}}\right]_n\mathrm{H} + n\ \mathrm{HCl}$$

Nylon 6-10

The chemical industry devotes much of its research activity and its manufacturing facilities to the production of polymers, yet polymerization reactions as laboratory exercises seldom are considered in the same perspective as the more classical preparations. There are several reasons for a lack of emphasis on polymerization reactions in the laboratory program. The reactions involved in the formation of a polymer in many cases are very complex and not as illustrative of organic reactions as are simpler preparations and starting materials required usually are not easily available. The laboratory preparation of a polyurethane foam, while representing no exception to these shortcomings, illustrates a poylmerization reaction of great commercial importance. The reaction is easy to carry out, rather spectacular to observe, and can be adapted to a very practical purpose.

In procedure (A) of this experiment a polyurethane foam is prepared from two substances of natural origin, castor oil and glycerol. Although commercial foams are not likely to be prepared from such simple starting materials, the principles behind the laboratory and commercial preparations are the same.

The simple urethanes may be prepared by the reaction of alcohols with isocyanates according to the following equation.[1]

$$\underset{\text{An isocyanate}}{R{-}N{=}C{=}O} + \underset{\text{An alcohol}}{R'OH} \rightarrow \underset{\text{A urethane}}{R{-}\overset{\overset{\displaystyle H}{|}}{N}{-}\overset{\overset{\displaystyle O}{\|}}{C}{-}OR'}$$

The polyurethanes are prepared from diisocyanates and polyhydric alcohols. Reactants with two or more functional groups are capable of polymerizing until all material is incorporated into one giant, cross-linked molecule. A small amount of water in the reaction mixture results in the production of sufficient unstable carbamic acids which, on decomposition, provide the carbon dioxide that causes the foam to rise.

$$R{-}N{=}C{=}O + H_2O \rightarrow \underset{\text{A substituted carbamic acid}}{R{-}\overset{\overset{\displaystyle H}{|}}{N}{-}\overset{\overset{\displaystyle O}{\|}}{C}{-}OH}$$

$$R{-}\overset{\overset{\displaystyle H}{|}}{N}{-}\overset{\overset{\displaystyle O}{\|}}{C}{-}OH \rightarrow R{-}NH_2 + CO_2$$

The process whereby a polyurethane foam is produced is somewhat like baking a cake. The small carbon dioxide bubbles fill the polymeric material with innumerable pores which make the foam "light." If these pores are closed the foam is a rigid one and the enmeshed gas is trapped to provide the remarkable insulating and buoyant properties for which this type of material is noted. The structural unit of a polyurethane is shown on the following page.

[1] Although no small molecule is liberated during polymer formation, these polymers are labeled condensation polymers because the polymer *could be considered* to result from the reaction of a carbamic acid and an alcohol with the loss of water.

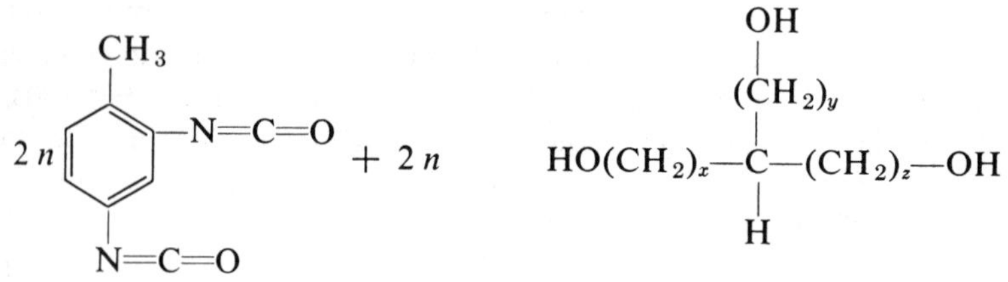

$$2n\ CH_3C_6H_3(N{=}C{=}O)_2 + 2n\ HO(CH_2)_x-\underset{\underset{H}{|}}{\overset{\overset{OH}{|}}{\overset{(CH_2)_y}{|}}}{C}-(CH_2)_z-OH$$

4-Methyl-*m*-phenylene diisocyanate (Tolyl diisocyanate, TDI)

A polyhydric alcohol[1]

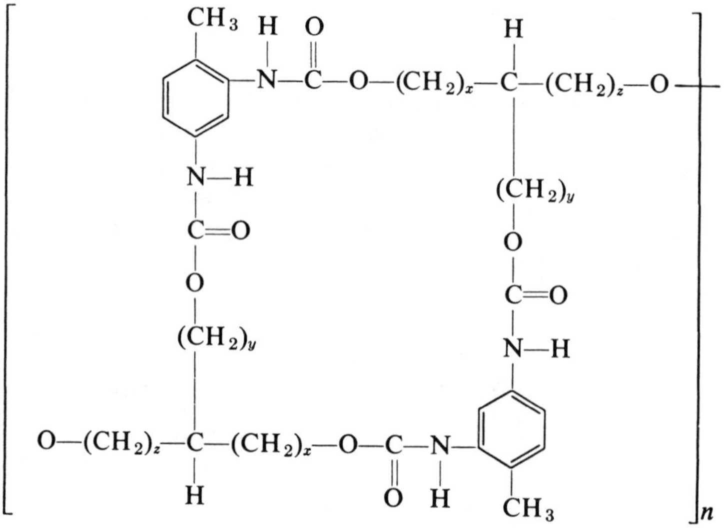

Structural unit of a polyurethane

[1]The molecular weight of a polyhydric alcohol used in a polyurethane foam may vary from 500 to 3000. The small letters *x*, *y*, and *z* therefore represent large numbers (10 to 70). While the alcohol in the sample equation is shown as a simple triol, commercial "Polyols" contain additional hydroxyl groups as well as ether linkages.

Procedure

A. THE PREPARATION OF A POLYURETHANE FOAM

In a plastic lined or waxed container (Note 1) weigh out 35 g of castor oil and 10 g of glycerol. To the combined reactants add 5 drops of stannous octoate, 10 drops of silicone oil (Dow-Corning 202), and 10 drops of water (use a dropper). Blend these ingredients thoroughly by stirring with a glass rod to a creamy, viscous mass. Next, weigh in a separate container (paper cup or beaker) 30 g of 4-methyl-*m*-phenylene diisocyanate (tolyl diisocyanate, TDI). **CAUTION!** *Avoid contact with skin, eyes, or clothing.* May cause skin irritation (Note 2). Add the correctly weighed amount of TDI *in one portion* to the previously mixed ingredients. Mix *rapidly* and *thoroughly* to a smooth, creamy, homogeneous mixture. When the container becomes warm to the touch and small bubbles begin to form, immediately discontinue stirring, place the container on a piece of newspaper *in a fume hood* and allow the reaction to proceed spontaneously. After foaming has ceased, allow the polymeric material to cool and "set" for 4 or 5 hours before removing the form. The volume of foam produced by the above formulation is approximately 75 cubic inches (Note 3). When the foam is fairly firm show it to your instructor, who may ask you to slice off a small section with a sharp knife to attach to your report.

Polyurethane foams produced from castor oil and glycerol usually will be of the compressible type and will set with varying degrees of shrinkage. Athough the practical applications of such foams are limited, the reaction is an excellent one to illustrate the polyfunctional requirements for the preparation of this type of elastomer.

Note 1. A paper milk carton (quart capacity) with the top removed or a paper cup (milk-shake size) serves very well as a container for the foaming reaction.

Note 2. TDI reacts vigorously with the moisture in air or in the skin to ultimately produce substituted ureas. Recap bottle promptly after using.

Note 3. Freshly formed foam often is sticky to the touch. Should any polyhydric alcohols be gotten on the workbench or hands accidentally, they may be removed with a 1:1 isopropyl alcohol-acetone mixture.

B. PREPARATION OF NYLON 6-10

Using a cotton swab (or your little finger) coat the inner surface of a 15 mm × 60 mm vial (a litmus paper vial will do nicely) with a thin layer of silicone oil (Note 1). Fill the vial almost to the half mark with a 5% solution of sebacoyl chloride in carbon tetrachloride. Hold the vial in an inclined position and *very slowly* add onto the carbon tetrachloride layer an equivalent volume of a 5% aqueous solution of hexamethylenediamine. A reaction product will form immediately at the interface of the two immiscible layers. Clamp the vial to a ring stand. Reach through the upper layer with a hook made from a 6-inch length of copper wire and draw up the Nylon which has formed (see Fig. 29.1). This will expose fresh reactants from both layers to produce additional polymer. Slowly and steadily continue to remove new material as one continuous thread. The thread may be wound onto a piece of cardboard as it is formed or it may be led into a 600-ml beaker filled with water. When the reactants have been reduced to about half of their initial volumes, stopper the vial and shake vigorously to thoroughly mix the remainder of the two layers. Transfer the resulting white opaque mass to a beaker and wash once with 95% alcohol and once with water. Squeeze the lump of material as dry as possible between paper towels and examine it. Does it have any tensile strength in this form? Transfer the dried nylon matting to a small test tube and try to melt it over a low flame. Does it melt easily? If you succeed in melting it, try drawing from the melt a thin fiber with the wire used in the first part of the experiment. If you are successful in drawing a fiber from the molten material, compare its strength with that of the thread produced from the two layers. Attach a short length of your "rope" to your report form.

Note 1. Coating the inner surface of the sample vial prevents nylon from sticking to the glass wall and insures a continuous thread.

FIGURE 29.1 Removing the Nylon thread from the interface of two immiscible layers.

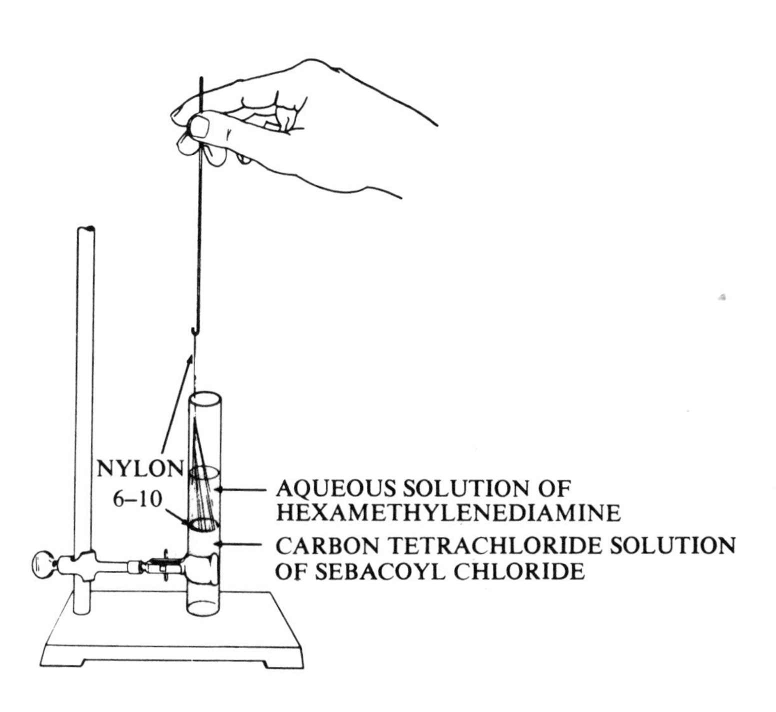

TEXT REFERENCE

Report: 29

Chapter Pages

Section Desk NAME

POLYMERIZATION

Attach a thin section of your polyurethane foam below.

Attach a short length of your nylon “rope” below.

Questions and Exercises

1. Castor oil may be considered to consist largely of the *glyceride* of ricinoleic acid.

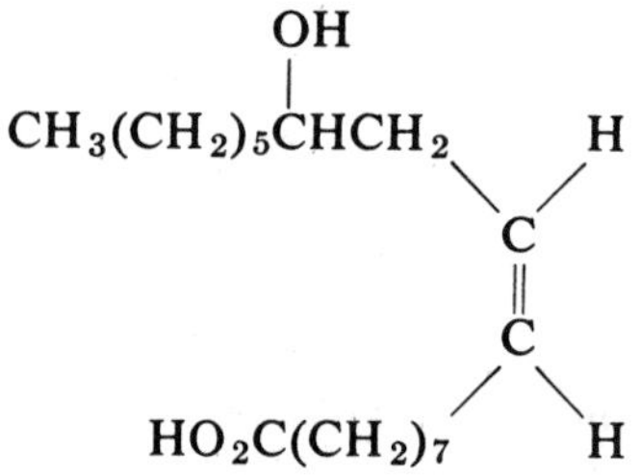

$$\begin{array}{ccc} \overset{\displaystyle OH}{\overset{|}{CH_3(CH_2)_5CHCH_2}} & & H \\ & \diagdown \quad \diagup & \\ & C & \\ & \| & \\ & C & \\ & \diagup \quad \diagdown & \\ HO_2C(CH_2)_7 & & H \end{array}$$

   Write the structure of the structural unit of your polyurethane foam based on (a) this glyceride and (b) glycerol.

2. Write the equations for the preparation of the following polymers: (a) Dacron, (b) Orlon, (c) Teflon, (d) polystyrene.

3. Define the term *copolymer*. Give an example (structural formula) of a copolymer.

4. The principal difference between *chain* and *step* polymerization is that in chain polymerization all of the intermediates are unstable species such as radicals, anions, or cations, whereas in step polymerization the intermediates are reasonably stable organic molecules. Illustrate this difference by writing the equations for the reactions of three molecules of ethylene by a free-radical chain process and equations for the reactions of three molecules of sebacoyl chloride with three molecules of hexamethylenediamine. Until very recently, these processes were called *addition* and *condensation* polymerization, respectively. Why do you suppose polymer chemists chose the new names above?

Experiment 30

Time: $3\frac{1}{2}$–4 hours

The Coupling of Aromatic Diazonium Compounds; Dyes and Dyeing

Nearly one-half of all synthetic dyes now in use are azo dyes prepared by **coupling** aromatic diazonium salts with phenols, naphthols, or aromatic amines.

$$p\text{-}O_2N\text{-}C_6H_4\text{-}NH_2 \xrightarrow{HCl} p\text{-}O_2N\text{-}C_6H_4\text{-}\overset{+}{N}H_3Cl^- \xrightarrow[0\text{–}5^\circ]{NaNO_2} p\text{-}O_2N\text{-}C_6H_4\text{-}\overset{+}{N}_2Cl^-$$

p-Nitrobenzene-diazonium chloride

$$p\text{-}O_2N\text{-}C_6H_4\text{-}\overset{+}{N}_2Cl^- + C_{10}H_7OH \rightarrow O_2N\text{-}C_6H_4\text{-}N{=}N\text{-}C_{10}H_6OH + HCl$$

β-Naphthol   Para red

The coupling reaction is carried out in alkaline, neutral, or in weakly acidic solution. Inasmuch as the diazotization and coupling reactions take place within the fabric at the temperature of an ice bath, azo dyes produced in this manner sometimes are referred to as "ice colors." One of the first dyes of this type to be prepared was **para red,** the preparation of which is part of the present experiment. Its use illustrates a method of dyeing known as **ingrain** dyeing. Two other common methods of dyeing — **mordant** and **direct** — also will be illustrated. It is very important that you exercise extreme care in dyeing operations. Avoid handling samples with the fingers; use glass rods. Avoid splashing or spilling. Wear your apron or some protective garment.

Procedure

A. AZO DYES AND INGRAIN DYEING. (PARA RED)

Place 1.0 g of *p*-nitroaniline in a 50-ml beaker; add 25 ml of water and 1 ml of concentrated hydrochloric acid, Place the beaker in an ice bath and cool to 0–5°. Add the to cold *p*-nitroaniline hydrochloride solution slowly and with stirring a cold (0–5°) solution of 1.0 g of sodium nitrite in 5 ml of water. While the diazonium salt solution is kept in the ice bath, prepare in a 250-ml beaker a solution of 1 g of β-naphthol in 100 ml of 2% sodium hydroxide. Immerse a piece of clean, cotton cloth or bandage gauze (1 inch × 2 inches) in the alkaline β-naphthol solution, squeeze out as much solution as possible, and hang the cloth up to dry. Dilute the cold diazonium salt solution with 200 ml of ice water and dip the dry, β-naphthol-soaked cloth in the solution of diazotized amine. Puddle the cloth with a glass rod, remove, and again air dry. Attach the dry, dyed cloth to your report form.

B. TRIPHENYLMETHANE DYES. MORDANT DYEING

Prepare the following solutions.

(1) Two dye baths, prepared from (a) 0.1 g malachite green in 200 ml of hot water and (b) 0.1 g methyl violet in 200 ml of hot water.

(2) A mordant bath prepared by dissolving 0.2 g of tannic acid in 150 ml of water.

(3) A fixing bath prepared by dissolving 0.2 g tartar emetic (potassium antimonyl tartrate, $KSbOC_4H_4O_6$) in 150 ml of water.

Each of these solutions must be heated to 80–90° (on the steam bath or over a burner flame) before use.

Soak four clean pieces of cotton cloth in the hot (80–90°) mordanting solution for one minute, squeeze as dry as possible, and transfer two pieces to the fixing bath. Allow the samples to remain in the fixing bath at 80–90° for 5 minutes. Remove the cloth samples from the fixing bath and transfer to the two separate dye baths. Allow each of the cloth samples to remain in the hot (90°) dye baths for about 5 minutes. Remove, wash with cold water, and hang them up to dry. Repeat the dyeing process with the two pieces of cotton cloth that have had the tannic acid treatment but have not had a fixing bath. Repeat the dyeing process using two pieces of cotton cloth that have had neither previous treatment. Which process produces a fast color? Attach all six samples to your report form.

C. NITRO DYES. DIRECT DYEING

Dissolve about 0.1 g of trinitrophenol (picric acid) in 250 ml of hot water. Immerse in the bath pieces of wool, silk, and cotton for 2–3 minutes. Remove the cloth, rinse, and dry. Attach the samples to your report form.

TEXT REFERENCE

Report: 30

Chapter Pages

Section Desk NAME

AZO DYES AND INGRAIN DYEING (PARA RED)

Cotton

TRIPHENYLMETHANE DYES AND MORDANT DYEING

Cotton (dye only)	Cotton (mordant and dye)	Cotton (mordant, fix, and dye)

NITRO DYES AND DIRECT DYEING

Cotton	Wool	Silk

Questions and Exercises

1. What substances other than tannic acid may be used as mordants?
2. Why is picric acid a good direct dye for silk and wool but not for cotton?
3. Name methods of dyeing other than the ones practiced in this experiment and give examples of dyes used in each.
4. With the aid of your text draw the structures of (a) indigo (b) Congo Red, and (c) malachite green. Enclose the chromophore with a broken line. Use a solid line to encircle the auxochromes in (a) and (c).
5. Write the equation for the probable method of synthesis of Congo Red (diazotization and coupling steps only).
6. In the red form, Congo Red may be considered to be a salt in which the —SO_3H groups have been converted to —SO_3^- groups (by treatment with base). In the blue form, two protons are added to the molecule to give two substances (differing principally in the location of the protons) in equilibrium with each other. One of these has the protons added to the —NH_2 groups to form —NH_3^+ groups. What is the probable site of protonation in the other substance (which is largely responsible for the blue color)? (It is *not* the SO_3^- group!) Write two resonance structures for the second protonated substance that will show the nature of the chromophore responsible for the blue color.

Experiment 31(a)

Time: $3\frac{1}{2}$ + 2 + 4 hours

*A Reaction Sequence: The Preparation of 4-*tert*-Butylbenzoic Acid*

This preparation illustrates two useful and general reactions: (1) the bromination of aromatic hydrocarbons, and (2) the carbonation of the Grignard reagent. It also gives the student opportunity to carry out a short reaction sequence. The sequence involves four successive steps as indicated by the following equations: (1) The preparation of 4-bromo-*tert*-butylbenzene, (2) the formation of 4-*tert*-butylphenylmagnesium bromide, (3) the carbonation of the Grignard reagent, and (4) the hydrolysis of the Grignard addition compound.

$$(1)\quad C_6H_5C(CH_3)_3 + Br_2 \xrightarrow{Fe} 4\text{-}Br\text{-}C_6H_4C(CH_3)_3$$

$$(2)\quad 4\text{-}Br\text{-}C_6H_4C(CH_3)_3 + Mg \xrightarrow[\text{ether}]{\text{Anhydrous}} 4\text{-}BrMg\text{-}C_6H_4C(CH_3)_3$$

$$(3)\quad 4\text{-}BrMg\text{-}C_6H_4C(CH_3)_3 + CO_2 \longrightarrow 4\text{-}(BrMg{-}O{-}\overset{O}{\overset{\|}{C}})\text{-}C_6H_4C(CH_3)_3$$

$$(4)\quad 4\text{-}(BrMg{-}O{-}\overset{O}{\overset{\|}{C}})\text{-}C_6H_4C(CH_3)_3 \xrightarrow{H_3O^+} 4\text{-}(HO{-}\overset{O}{\overset{\|}{C}})\text{-}C_6H_4C(CH_3)_3$$

The continuity of the sequence may be interrupted after step (1), or (3). Once the Grignard reagent is formed, it should be carbonated without delay. After the Grignard reagent has been carbonated, it may be allowed to stand until the next laboratory period.

PART I PREPARATION OF 4-BROMO-*tert*-BUTYLBENZENE

Procedure

Assemble an apparatus consisting of a clean, dry 250-ml round-bottomed flask and a reflux condenser fitted with one of the gas (hydrogen bromide) absorption devices shown in Fig. 31.1. One product of the reaction is hydrogen bromide, an acidic gas, and unless the reaction is carried out in a fume hood, the gas must be absorbed. After your apparatus has been approved by your instructor, remove the reaction flask and place in it 26.8 g (0.2 mole, 31 ml) of *tert*-butylbenzene. Add 24 g (0.15 mole, 8.2 ml) of bromine from the bromine buret (**HOOD!**) directly to the *tert*-butylbenzene (Note 1). Refit the condenser and circulate water *slowly* through the water jacket. Have ready a pan of crushed ice and water to arrest the reaction should it become too vigorous. Through the open top of the condenser introduce five #3 tacks (Note 2). Connect the gas trap to the condenser. Reaction frequently is immediate, as will be indicated by the rising of hydrogen bromide bubbles through the surface of the mixture. Should the reaction become too lively, immerse the flask in the ice water without delay. The reaction, if too rapid, will cause bromine vapor to be carried over into the gas trap. The reaction should continue at a satisfactory rate without the necessity of continued cooling and may be considered complete when the space above the surface of the reaction mixture is clear of bromine vapor and bubbling at the surface ceases. (Approximately 45 minutes.) Heat the reaction mixture in a warm water bath (75°) for 10–15 minutes. Remove the gas trap and through the top of the condenser add 75 ml of cold water. Remove the condenser, transfer the reaction mixture to a separatory funnel, and separate the layers. Discard the aqueous layer. Wash the organic layer a second time with 75 ml of 5% sodium carbonate solution. (**CAUTION!** *Effervescence.*) Separate, and wash a third time with 50 ml of water (Note 3). Draw off and discard the wash water. Dry the organic layer over calcium chloride (Note 4). While your product is drying set up the apparatus required to carry out the procedure described under Part II (Note 5).

Note 1. Exercise caution in handling bromine. To avoid contact with liquid bromine or bromine vapor it is advisable to wear plastic gloves (see Appendix, Chemicals and Reagents Required, Experiment 17). Should you accidently spill bromine on your skin, apply copious quantities of glycerol, wash with water, and apply a slurry of sodium bicarbonate.

Note 2. Old, used tacks serve as an excellent catalyst for this reaction. If new, blued tacks are used, they should be cleaned by immersing in dilute nitric acid for 1–2 minutes or filed to brightness.

Note 3. If your organic layer is orange or brown in color due to the presence of unreacted bromine, add 10 ml of a saturated solution of sodium

FIGURE 31.1 Devices for the absorption or removal of acidic gases during refluxing.

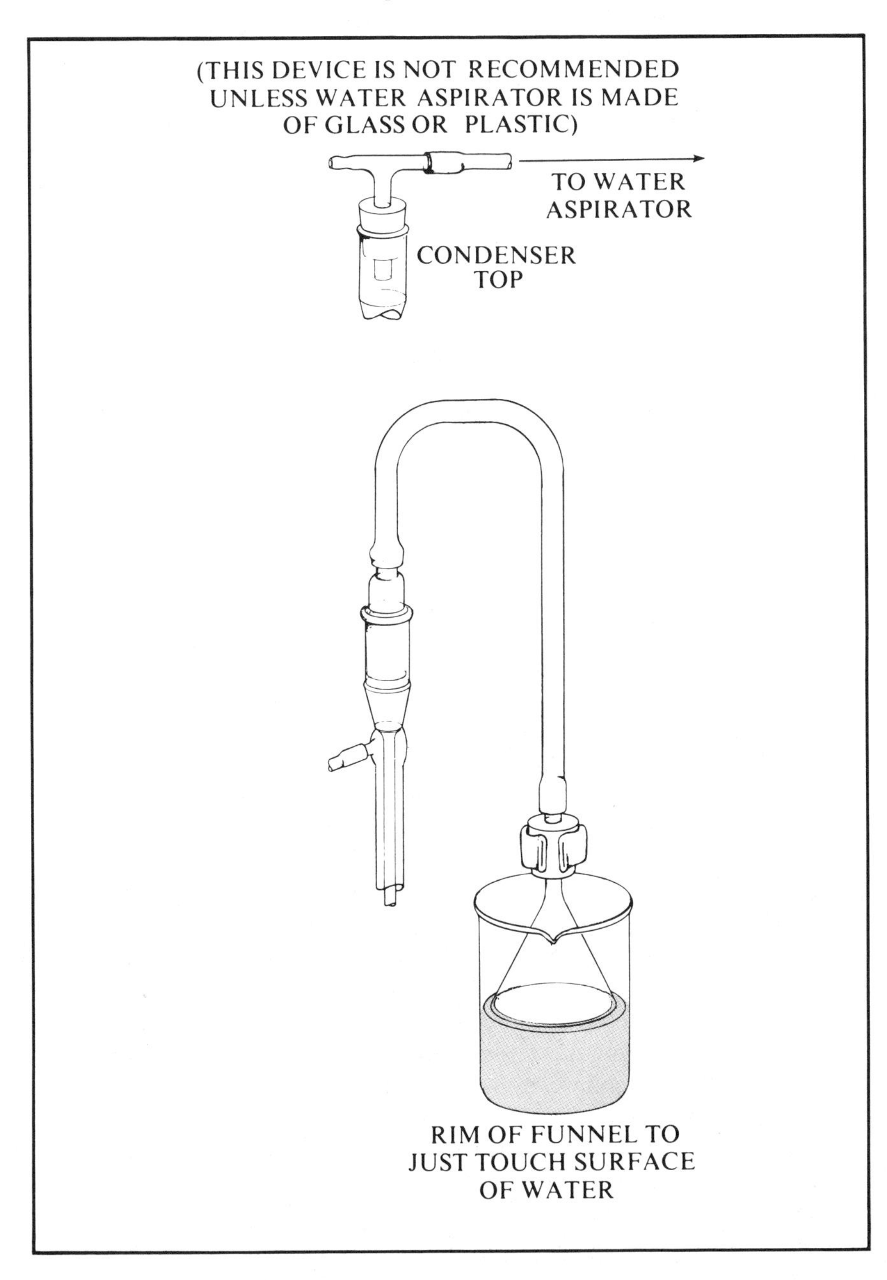

bisulfite in water to the 50 ml of water used for the third wash, and wash the organic layer a fourth time with another 50-ml portion of water.

Note 4. The sequence may be interrupted at this stage and optional Experiment 31(b) performed to determine the percentage yield.

Note 5. The high boiling point of *tert*-butylbenzene (169°) and that of 4-bromo-*tert*-butylbenzene (232°) preclude an easy separation of the reaction product by any means other than a fractional distillation under diminished pressure. The entire reaction mixture containing approximately 24 g (75% yield) of 4-bromo-*tert*-butylbenzene is used in the preparation of the Grignard reagent. Any unreacted *tert*-butylbenzene will remain in the solvent ether. Should the instructor prefer a separation of 4-bromo-*tert*-butylbenzene, it may be distilled at 80–81°/2 mm. Yield 75–78%.

PART II THE PREPARATION OF A GRIGNARD REAGENT IN THE 4-*tert*-BUTYLBENZOIC ACID SEQUENCE

The preparation of 4-*tert*-butylphenylmagnesium bromide is accomplished by following a procedure only slightly modified from that described in Experiment 21 for the preparation of *n*-butylmagnesium bromide.

Procedure

Assemble an apparatus like that illustrated in Figure 21.1 using clean, *dry* glassware and one ring stand. Place 2.64 g (0.11 mole) of magnesium turnings in the flask, wet the turnings with 2–3 ml of 4-bromo-*tert*-butylbenzene, then add 15 ml of anhydrous ether, and a small (pin head size) crystal of iodine in the flask. Into a dropping funnel place the remainder of the reaction mixture and an additional 35 ml of anhydrous ether. Swirl the mixture to insure complete solution. Add at one time approximately half of the ether solution of 4-bromo-*tert*-butylbenzene to the magnesium turnings. Warm the mixture to a gentle reflux on a 50° water bath. You will know that the reaction has begun when the pale amber color of the solution fades and becomes an opalescent white and refluxing of the ether becomes spontaneous without further heating. Add the remainder of the reagent from the dropping funnel dropwise to maintain a fairly rapid reflux. Cooling of the refluxing mixture is unnecessary unless violent boiling and frothing occur and ether vapor escapes from the top of the condenser. After the reaction is well on its way, a darkening of the mixture may be observed. When refluxing subsides and nearly all of the magnesium has been consumed, immerse the reaction mixture in a 50° water bath and allow it to reflux for an additional 15 minutes. Remove the condenser from the flask and pour the entire contents slowly but steadily onto 35 g of crushed "dry ice" contained in a 600-ml beaker, taking care to keep unreacted magnesium in the flask (Note 1). Cover the reaction mixture with a watch glass and let stand until the excess dry ice has completely sublimed. The Grignard addition compound will appear as a viscous glassy mass. If the mass is too viscous to stir, add an additional 50 ml of ether (Note 2).

Hydrolyze the Grignard addition product by adding to it 50 g of crushed ice to which 15 ml of concentrated hydrochloric acid has been added. Stir the mixture until two layers appear, then transfer the mixture to a separatory funnel. Draw off the lower aqueous layer and discard it. Wash the upper ether layer once with 50 ml of water (Note 3); then, separate and discard the wash water layer. Extract the ether layer three times with 50-ml portions of 5% sodium hydroxide solution. Combine the basic extracts and discard the ether layer. Wash the combined basic extracts once with 50 ml of ether. Discard the ether layer. Place the combined basic extracts in a 400-ml beaker and set in a boiling water bath **(HOOD! Extinguish all flames!)** and stir until the ether dissolved in the alkaline solution has been boiled out (Note 4). Cool the alkaline solution in an ice-water bath and precipitate the 4-*tert*-butylbenzoic acid by the addition of 20 ml of concentrated hydrochloric acid. Collect the precipitated acid on a Büchner funnel by suction filtration. Wash the collected crystals with several small portions of cold water and dry. The product is fairly pure as collected, but a 0.5-g sample may be recrystallized from dilute alcohol for a melting point determination. Weigh the product and turn it in with your report. Yield approximately 11 g (56%), mp 164°.

Note 1. Exercise caution in handling dry ice. Contact with the skin can cause severe burns or frostbite. Always use cotton gloves or tongs. The dry ice is best broken up by wrapping one lump in a clean, dry towel and beating it on the desk top or floor.

Note 2. The reaction sequence may be interrupted at this stage and completed in the next laboratory period.

Note 3. If the ether layer appears dark yellow or brown due to the presence of iodine, add 10 ml of a saturated solution of sodium bisulfite in water to the 50 ml of wash water.

Note 4. Ether is soluble in water to the extent of 7%. Unless the ether is removed before the substituted benzoic acid is precipitated, the product may appear as a waxy solid and not crystalline.

TEXT REFERENCE

Report: 31(a)

Chapter Pages

Section Desk NAME

PREPARATION OF 4-*tert*-BUTYLBENZOIC ACID

Reaction equation

4-Bromo-*tert*-butylbenzene ($BrC_6H_4C(CH_3)_3$) $\xrightarrow{\text{1. Mg, ether; 2. } CO_2\text{; 3. } H_3O^+}$ 4-*tert*-Butylbenzoic acid ($HOOCC_6H_4C(CH_3)_3$)

	4-Bromo-*tert*-butylbenzene	4-*tert*-Butylbenzoic acid
Quantities	24 g*	______
Mol. Wt.	______	______
Moles	______	______

Theoretical yield ______

Actual yield ______g

Percentage yield ______

mp ______°C

*Based on an assumed 75% yield of 4-bromo-*tert*-butylbenzene.

Questions and Exercises

1. What is the function of the iron tacks used in the preparation of 4-bromo-*tert*-butylbenzene? Write a mechanism for the bromination of *tert*-butylbenzene that supports your answer. What other catalysts could be substituted for the tacks?
2. What is the function of anhydrous ether in the preparation of a Grignard reagent? In some advanced work, in addition to the use of dry ether, a slow stream of dry nitrogen is passed through the vessel in which a Grignard reagent is being prepared or used. From what is the Grignard reagent being protected by the dry nitrogen? Write equations for two or three reactions that could take place to destroy all or part of the Grignard reagent if it were not protected.
3. Two easily available forms of carbon dioxide are dry ice and the compressed gas. Dry ice has the obvious advantage of being easy to use in the carbonation of Grignard reagents. What is an obvious disadvantage of the use of this form of carbon dioxide? What can be done to minimize the incidence of this side reaction?
4. What products would you expect from the addition of phenylmagnesium bromide to (a) methyl ethyl ketone, (b) methyl acetate, and (c) acetonitrile?

Experiment 31(b)

(Optional)

The Gas Chromatographic Separation and Identification of the Reaction Product Resulting from the Bromination of tert-*Butylbenzene*

An interesting departure from the sequence of steps comprising Experiment 31 is the gas chromatographic analysis of the reaction mixture that results from the bromination of *tert*-butylbenzene. An approximate percentage yield of 4-bromo-*tert*-butylbenzene may be determined via GLC and used as a basis for calculation of the final yield of 4-*tert*-butylbenzoic acid.

Procedure

The procedure that follows is essentially the same as that described in Experiment 6, Part IV, except for the following changes in instrument adjustment. An optimum flow rate of about 20 ml/min should be set for the carrier gas (Note 1), and a temperature of 235° should be set for the injection port. Start the recorder and set the attenuation at a sensitivity value that ensures a good measureable signal. A setting of 8× or 16× is satisfactory on the Gow-Mac Model 69-100 instrument. Draw a 0.5-μl sample of the unrectified but dried reaction mixture into the syringe and follow this liquid sample by drawing in an additional 5 μl of air. Insert the needle into the septum, being careful to avoid contact with the hot injection port.

The order of appearance of the components in the effluent gases will correspond to the order of their boiling points—that is, *tert*-butylbenzene will be eluted first. Total elution time is 3 minutes.

Identify each peak on the chart paper and indicate the attenuator setting at which each signal was obtained. Determine the areas of the *tert*-butylbenzene and 4-bromo-*tert*-butylbenzene peaks (see Experiment 6, p. 58) and calculate from the total areas the percentage of each component in the mixture of undistilled product. Record the percentages of each component in the mixture on your report form and attach the chromatograph.

Note 1. The column employed in the chromatography apparatus used in this experiment was a 4-foot, ¼-inch column containing Chromosorb DC-200.

TEXT REFERENCE

Report: 31(b)

Chapter Pages

Section Desk NAME

PREPARATION OF 4-BROMO-*tert*-BUTYLBENZENE

Reaction equation

$$C_6H_5C(CH_3)_3 + Br_2 \xrightarrow{Fe} \text{4-}BrC_6H_4C(CH_3)_3$$

tert-Butylbenzene 4-Bromo-*tert*-butylbenzene

Quantities 31 ml (26.8 g)

Mol. Wt. ________ ________ ________

Moles ________ ________ ________

Theoretical yield________g

Percentage yield________(by GLC)

Experiment 32

Time: 2 × 3 hours

Sulfanilamide

The preparation of sulfanilamide illustrates the synthesis of an important medicinal product. The reaction sequence which leads to the preparation of this "sulfa" drug illustrates the sulfonation of an aromatic ring bearing a "protected" amino group.

$$\underset{\text{I}}{C_6H_5NHCOCH_3} \xrightarrow{HOSO_2Cl} \underset{\text{II}}{p\text{-}CH_3CONHC_6H_4SO_2Cl} \xrightarrow{NH_3} \underset{\text{III}}{p\text{-}CH_3CONHC_6H_4SO_2NH_2} \xrightarrow[HCl]{H_2O} \underset{\text{IV}}{p\text{-}NH_2C_6H_4SO_2NH_2}$$

Acetanilide (I) may be readily chlorosulfonated with chlorosulfonic acid to produce *p*-acetamidobenzenesulfonyl chloride (II). The sulfonyl chloride, when treated with ammonia, yields the sulfonamide, *p*-acetamidobenzenesulfonamide (III). When (III) is treated with aqueous hydrochloric acid only the carboxylic acid amide is hydrolyzed to produce *p*-aminobenzenesulfonamide (IV). *p*-Aminobenzenesulfonamide usually is called sulfanilamide because the common name of the parent acid is sulfanilic acid.

Procedure

The preparation of sulfanilamide is accomplished in three procedural steps. Steps (A) and (B) of the sequence must be completed in the same laboratory period. Step (C) may be deferred until the next laboratory period.

CAUTION! *Chlorosulfonic acid is an extremely noxious chemical and should be handled with care. Use only dry graduates and flasks with this reagent. Should you spill chlorosulfonic acid on your skin, wash it off immediately with water.* **Wear your safety glasses! Perform parts A and B in a good exhaust hood!**

A. PREPARATION OF *p*-ACETAMIDOBENZENESULFONYL CHLORIDE

In a dry 50-ml Erlenmeyer flask place 12.5 ml (22.5 g, 0.21 mole) of chlorosulfonic acid. Cool the acid to 10–15° in a bath of ice water. *Clamp the flask in place.* Add 5 g of finely powdered, dry acetanilide in small portions (about 0.5 g each), using a stirring rod to mix the ingredients and keeping the temperature in the range of 10–20°. After all or most of the acetanilide has dissolved, remove the flask from the ice bath and clamp it in place on a water bath. Heat the solution gently on the water bath for 20 minutes to complete the reaction. Pour the reaction mixture slowly and carefully (avoid spattering) into a mixture of 100 g of ice and 50 ml of water. Rinse the flask with cold water and add the water to the main reaction mixture. Stir the mixture well, breaking up any lumps with a stirring rod. Filter the product on the Büchner funnel with suction and wash the product with 50 ml of cold water. Use the product immediately in the next step.

B. PREPARATION OF *p*-ACETAMIDOBENZENESULFONAMIDE

Transfer the crude, damp product from procedure (A) to a 125-ml Erlenmeyer flask and add 20 ml of concentrated ammonium hydroxide, mixing well with a stirring rod. A reaction usually begins immediately and the mixture becomes warm. Incorporate the flask in a reflux set-up and heat the mixture on a water bath for 15 minutes. Remove the flask and place it in an ice bath. When the mixture is cold, add 40–50 ml of dilute (6 *N*) hydrochloric acid until the mixture is acid to litmus paper. Continue cooling in the ice bath until the mixture is thoroughly cold, and filter the product on the Büchner funnel with suction. Wash the product with 50 ml of cold water. The product may be used directly in the next step or allowed to dry in air.

C. PREPARATION OF *p*-AMINOBENZENESULFONAMIDE

Transfer the crude product from procedure (B) to a 125-ml flask and add a mixture of 10 ml of water and 5 ml of concentrated hydrochloric acid. Incorporate the flask in a reflux assembly and heat (gently at first to prevent charring) the mixture to boiling on a wire gauze over a low flame. Boil the solution for 20 minutes, during which time the solid should dissolve. Dilute the solution with an equal volume of water and, if discolored, add a small amount (match head size) of vegetable charcoal. Heat the solution to boiling and filter through a fluted filter into a 400-ml beaker. Add solid sodium carbonate in small portions with stirring until the solution is just alkaline to litmus. Chill the mixture thoroughly in the ice bath and collect the precipitate by suction filtration on the Büchner funnel. Wash the crystals with 25 ml of cold water and dry as much as possible on the suction filter. Recrystallize from water using about 10–12 ml of water per gram of crude product. Dry your product in the air and determine its melting point. Yield approximately 2 g. Submit your product with your report.

TEXT REFERENCE

Report: 32

Chapter Pages

Section Desk NAME ______________________________

SULFANILAMIDE

Reaction equation

$$C_6H_5\text{—}NH\text{—}C(=O)\text{—}CH_3 \xrightarrow[\text{3. } H_3O^+]{\text{1. } HOSO_2Cl \quad \text{2. } NH_3} H_2N\text{—}C_6H_4\text{—}SO_2NH_2$$

Acetanilide Sulfanilamide

	Acetanilide	Sulfanilamide
Quantities	5.0 g	________g
Mol. Wt.	________	________
Moles	________	________

Theoretical yield ________g

Actual yield ________g

Percentage yield ________ (Based on acetanilide)

mp ________°C

Questions and Exercises

1. Write equations for the preparation of (a) *p*-toluenesulfonyl chloride from toluene, (b) N-methylbenzenesulfonamide from benzenesulfonyl chloride.

2. Based on your equation from Exercise 1, why was it necessary to protect the amino group in your preparation of sulfanilamide? What sort of product might result if the amino group were not protected?

3. In the hydrolysis step in the sulfanilamide preparation procedure, why is the carboxamide hydrolyzed rather than the sulfonamide?

* 4. The ir and nmr spectra of sulfanilamide appear below. Identify in the ir spectrum the type of bond (and its mode of vibration) responsible for each of the principal absorption bands shown. In the nmr spectrum identify the protons responsible for each of the signals shown.

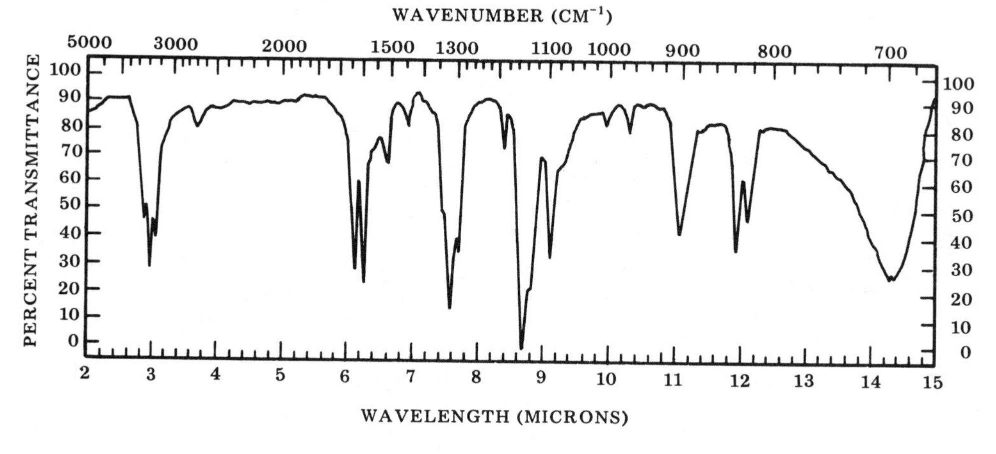

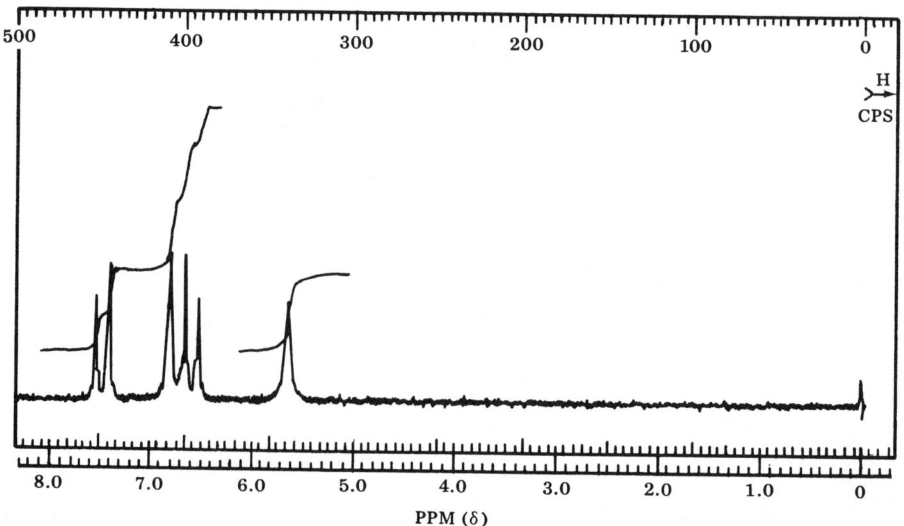

Experiment 33

Time: $2\frac{1}{2}$–3 hours

The Preparation of Pinacol Hydrate

The following experiments are related reaction sequences inasmuch as each begins with the bimolecular reduction of acetone to pinacol hexahydrate. Sequence I requires the preliminary drying of pinacol hydrate to the anhydrous form before it can be dehydrated by mild acid treatment to 2,3-dimethyl-1,3-butadiene. With the diene on hand, the stage is set for carrying out the Diels-Alder reaction by combining it with maleic anhydride to produce *cis*-4,5-dimethyl-1,2,3,6-tetrahydrophthalic anhydride. Sequence II begins with a change in the acid treatment above in order to favor a molecular rearrangement of pinacol to pinacolone over diene formation. A methyl ketone such as is exemplified by pinacolone suggests a route to pivalic acid *via* the haloform reaction. The two sequences may be shown in reaction form as follows:

$$(CH_3)_2C{=}O \xrightarrow{Mg(Hg)} \left[\begin{array}{c}(CH_3)_2\dot{C}{-}O\\ \\ (CH_3)_2\dot{C}{-}O\end{array}\!\!\!>Mg\right] \rightarrow \begin{array}{c}(CH_3)_2C{-}O\\ | \\ (CH_3)_2C{-}O\end{array}\!\!\!>Mg$$

Magnesium pinacolate

$$\xrightarrow{8\,H_2O} \begin{array}{c}(CH_3)_2C{-}OH\\ | \\ (CH_3)_2C{-}OH\end{array} \cdot 6\,H_2O$$

Pinacol hexahydrate

Sequence I:

$$\begin{array}{c}(CH_3)_2C{-}OH\\ | \\ (CH_3)_2C{-}OH\end{array} \cdot 6\,H_2O \xrightarrow{-6\,H_2O} \begin{array}{c}(CH_3)_2C{-}OH\\ | \\ (CH_3)_2C{-}OH\end{array} \xrightarrow{HBr} CH_2{=}C(CH_3){-}C(CH_3){=}CH_2$$

Pinacol hexahydrate → Pinacol → 2,3-Dimethyl-1,3-butadiene

$$\xrightarrow{\text{Maleic anhydride}}$$ *cis*-4,5-Dimethyl-1,2,3,6-tetrahydrophthalic anhydride

Sequence II:

$$\text{Pinacol hexahydrate} \xrightarrow{H_2SO_4} (CH_3)_3C{-}\overset{\overset{O}{\|}}{C}{-}CH_3$$

Pinacolone

$$\xrightarrow{(1)\ NaOH,\ Br_2;\ (2)\ H_3O^+} (CH_3)_3C{-}COOH$$

Pivalic acid

II I

Each multistep sequence requires the practice of careful laboratory techniques and attention to detail but affords an opportunity for especially interested students and those of sufficient attainment to repeat some of the important accomplishments of earlier organic chemists.

The reactions show five different chemical changes and illustrate (1) a bimolecular reduction of a ketone to a diol, (2) the dehydration of a diol to a diene, (3) the Diels-Alder [4 + 2] cycloaddition, (4) the formation of a carbocation with a resulting molecular rearrangement, and finally (5) an oxidative degradation *via* the haloform reaction. Much interesting chemistry is involved here, but the following experiments will serve better to illustrate organic reactions and to provide instructional opportunity than to demonstrate any facility with which one organic compound may be converted to another. Unfortunately, the yields resulting in each of the above reactions are low, but by carrying out the sequences as a team involving two or more students, yields may be combined and good positive results obtained from each procedure.

PRECAUTIONARY NOTE

Good laboratory practice requires that all potentially hazardous byproducts of a reaction be properly disposed of. Most chemistry laboratories have made arrangements for the disposal of wastes containing mercury or mercury salts. The student must follow the directions provided by the instructor for such disposal. In the event that facilities are not available for handling mercury wastes, Experiment 33 should be skipped and the reaction sequence should be started with Experiment 34 using commercially available pinacol hexahydrate.

Procedure

Assemble on a single ring stand a *dry* 500-ml round-bottomed flask fitted with a reflux condenser bearing a drying tube and a funnel (see Figure 21.1). Place in the flask 100 ml of anhydrous tetrahydrofuran[1] and dissolve in it 10 g of mercuric chloride. (**CAUTION!** *Mercuric chloride is very toxic. Handle with extreme care. Avoid contact with skin; avoid inhalation of any dust. Spills should be cleaned up at once using copious amounts of water.*) After the mercuric chloride has dissolved, add to the flask 12 g (0.5 gram atom) of bright magnesium

[1] **Note to Instructor.** When exposed to air repeatedly as in the case of old, nearly empty bottles of the reagent, tetrahydrofuran readily forms peroxides and may pose an explosive hazard if old tetrahydrofuran is used as starting material. A fresh bottle of tetrahydrofuran that has been stabilized should be used for this experiment. The presence of peroxides may be tested for with a 5% solution of potassium iodide acidified with a drop of sulfuric acid. This treatment, followed by a few drops of starch solution, produces a blue color if peroxides are present.

turnings. Swirl the flask. Next add by way of the funnel 75 ml of anhydrous acetone in 5–10 ml increments (Note 1). If the glassware and reagents used are dry, reaction is immediate and lively and will need to be moderated by periodic cooling of the flask in an ice-water bath. The reaction will proceed spontaneously for the next 30–45 minutes and should be allowed to continue as vigorously as possible without any undue amount of cooling. Allow refluxing to continue until it subsides of its own accord; then, using a steam bath or a heating mantle, reflux for an additional hour. The reaction mixture becomes progressively more viscous and turns into a gelatinous mass by the end of the reflux period. Stirring is difficult, but attempts should be made from time to time to break up the mass by vigorous swirling of the entire assembly. Hydrolyze the magnesium pinacolate by adding 150 ml of saturated sodium chloride solution (brine) and heat and swirl for 30 minutes. Allow the mixture to settle, then carefully decant the supernatant solution into an Erlenmeyer flask and set aside. Add 50 ml of ordinary tetrahydrofuran and 50 ml of brine to the grey precipitate which remains in the flask and heat for 15 minutes. Again, decant this extract and add it to the main solution. Filter the residue on a Büchner funnel and add the filtrate to the combined solutions obtained by decantation (Note 2). Place the residue in a container especially reserved for toxic wastes. Separate the organic layer from the brine. Discard the brine and reduce the volume of the combined solutions by removing approximately 100 ml of solvent by distillation. Precipitate the pinacol hydrate from the residual solution by adding 150 ml of petroleum ether and cooling in an ice-water bath. (**CAUTION!** *Petroleum ether is a very flammable solvent with a low flash point*). Collect the product on a Büchner funnel and press the crystals as dry as possible between layers of filter paper. The product is sufficiently pure to be used in subsequent preparations. Store in a tightly stoppered container (Note 3). Yield 16 g; mp 46–47°.

Note 1. The glassware and reagents used in this preparation must be *dry*, as in the preparation of a Grignard reagent.

Note 2. Filtration is very slow but may be expedited by the use of Celite or Filter-aid.

Note 3. Pinacol hydrate sublimes readily.

TEXT REFERENCE

Report: 33

Chapter Pages

Section Desk NAME

THE PREPARATION OF PINACOL HEXAHYDRATE

Reaction equation

$$2\ (CH_3)_2C{=}O + Mg(Hg) + 8\ H_2O \rightarrow$$

Acetone

$$\begin{matrix}(CH_3)_2C{-}OH \\ | \\ (CH_3)_2C{-}OH\end{matrix} \cdot 6\ H_2O + Mg(OH)_2$$

Pinacol

	Acetone	Magnesium	Pinacol hydrate
Quantities	______	______	______
Mol. Wt.	______	______	______
Moles	______	0.5 gram-atom	______

Theoretical yield ______ g (based on gram-atoms of Mg taken)

Actual yield ______ g

Percentage yield ______

Questions and Exercises

1. Al(Hg) may be used in place of Mg(Hg) to effect a bimolecular reduction of acetone. Draw the structure for aluminum pinacolate.
2. Draw the structure of, and name the product that results from, the bimolecular reduction of acetophenone.
3. When the compound of Question 2 is treated with sulfuric acid, two products are possible, but one predominates. Which is the major product and why is its formation favored?
4. Benzophenone is reduced to benzopinacol when a solution in 2-propanol is exposed to sunlight. Show how this reduction takes place by way of a free-radical mechanism.
5. What different products could result if 1-phenyl-1,2-propanediol were to undergo a pinacolic rearrangement? What factors determine which product is the likely one?

Experiment 34

Time: 3–4 hours

The Diels-Alder Reaction: The Preparation of cis-4,5-Dimethyl-1,2,3,6-tetrahydrophthalic Anhydride

One of the most important reactions of conjugated dienes is the Diels-Alder reaction—a [4 + 2]-cycloaddition that these unsaturated compounds undergo when reacted with active ethylenic or acetylenic compounds.

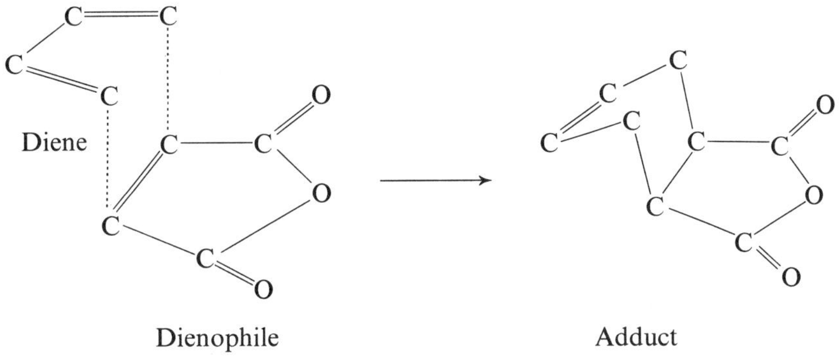

Diene

Dienophile

Adduct

The reaction involves a conjugated *diene* and a *dienophile* (Greek, diene loving) to yield a cyclohexene derivative. No catalyst is required, but the reaction is facilitated when the dienophile has an electron-withdrawing group conjugated with a multiple bond such as is found in an α,β-unsaturated carbonyl or nitrile. The reaction is stereospecific in that the configurations of both the diene and the dienophile are retained in the adduct.

In the present experiment we will illustrate the Diels-Alder reaction by combining 2,3-dimethyl-1,3-butadiene with maleic anhydride to give the cyclic adduct, *cis*-4,5-dimethyl-1,2,3,6-tetrahydrophthalic anhydride.

THE PREPARATION OF ANHYDROUS PINACOL

The water of hydration must be removed first before pinacol hydrate can be converted into 2,3-dimethyl-1,3-butadiene. The dehydration step is accomplished by refluxing the hydrate with an organic solvent that forms an azeotrope with water. Although pinacol hydrate is but sparingly soluble in cyclohexane, water is even less so. By heating the hydrate with cyclohexane under reflux and utilizing a water trap in our assembly as illustrated in Figure 34.1 we can effectively dehydrate the pinacol. As the azeotrope reaches the condenser, the solubility of water in cold cyclohexane is but a very small fraction of its solubility in hot cyclohexane, and water separates as a second phase, dropping to the bottom part of the water trap. Cyclohexane fills the upper portion of the trap and overflows back into the distilling flask. A calibration of the water trap permits a close approximation of the completeness of dehydration if the amount of material to be stripped of water is first weighed. However, the calibration feature, while convenient, is not necessary because it usually is quite obvious when droplets of water no longer form in the trap.

Procedure

Assemble a reflux apparatus with a Dean-Stark water trap interposed between a 250-ml distilling flask and the condenser, or, if this specially designed trap is not available, improvise a water trap by utilizing connecting adapters and a 25-ml flask as shown in Figure 34.1(a) (Note 1).

Weigh and place the entire yield of pinacol hydrate obtained from Experiment 33 into the distilling flask, add 100 ml of cyclohexane (**CAUTION!** *Flammable solvent. Low flash point!*), and heat to reflux using a heating mantle or a steam bath. **(NO FLAMES.)** Reflux until approximately 0.5 ml of water is collected for each gram of pinacol hydrate used. When no more water appears to pass over, discontinue refluxing, disconnect the water trap, and discard the water and cyclohexane collected. Rearrange your apparatus for distillation and distill again using a heating mantle or a steam bath. After removing the cyclohexane (bp 81°C) change receivers, and, *if others near you no longer are distilling cyclohexane*, use a Bunsen burner adjusted to a low flame to collect all product boiling between 170–175°C as anhydrous pinacol. Upon cooling, anhydrous pinacol will solidify (mp 38°C).

Note 1. To avoid any leakage at joint (x) in Fig. 34.1(a), the clamp supporting the condenser should be closed only tightly enough to keep the condenser in place yet allow its weight to bear upon the connecting adapter below and indirectly upon the water-receiving flask. In addition, a rubber band looped around the adapter above the side arm and around the clamp supporting the water receiver will help prevent leakage. All joints should be *lightly* greased.

FIGURE 34.1 Water separator assemblies. (a) Improvised trap from standard taper ware; (b) Dean-Stark trap.

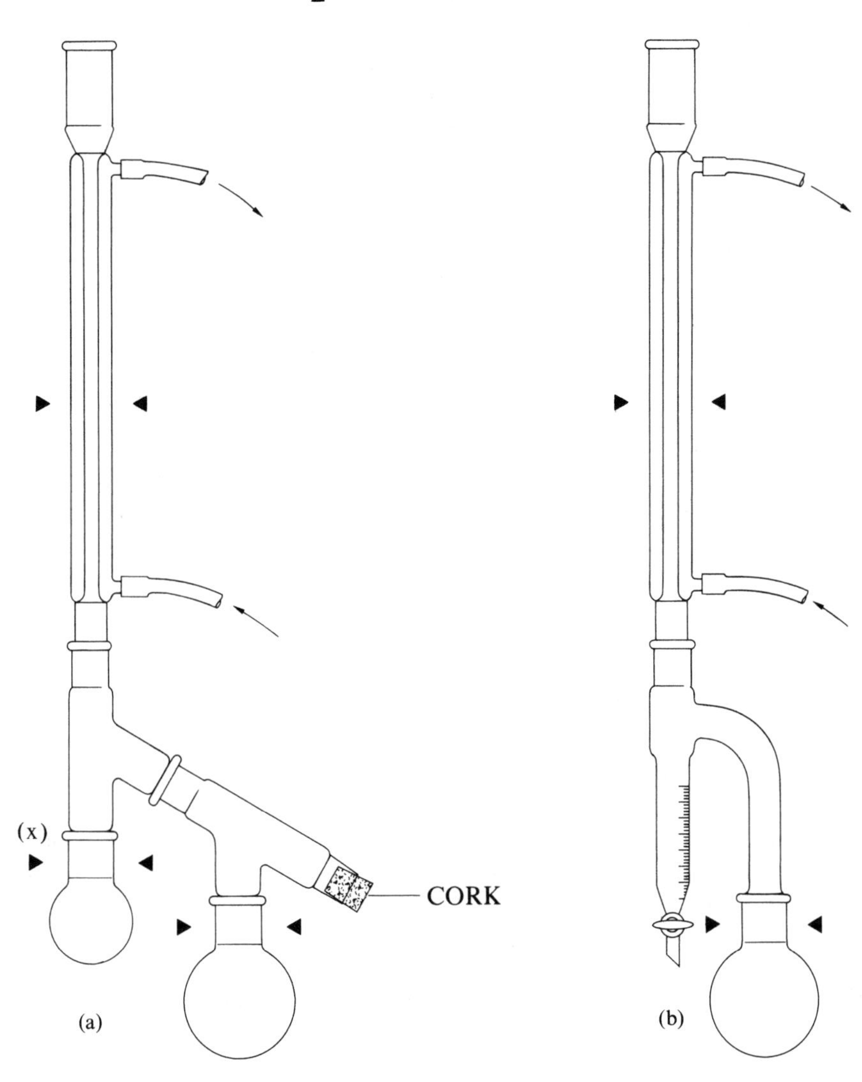

PART I THE PREPARATION OF 2,3-DIMETHYL-1,3-BUTADIENE

Assemble an apparatus for a small-scale fractional distillation like that illustrated in Figure 34.2 using a 50-ml round-bottomed flask as the boiling flask and a 25-ml flask as the receiver. Improvise a short, bead-packed fractionating column from a drying tube or from a modified Claisen adapter. Place 11.8 g (0.10 mole) of anhydrous pinacol and 1.0 ml of 48% hydrobromic acid in the flask. Add a "Boileezer" and distill using a steam bath or heating mantle. Insulate the packed column with aluminum foil to prevent too much heat loss. Collect all distillate that passes over below 90°C. The diene has a boiling point of 76°C, and pinacolone, some of which is also formed, has a boiling point of 106°C. Continue distillation until the temperature cannot easily be kept below 90°C even though a small volume (*ca.* 1–2 ml) remains in the distillation flask. Place the distillate in a small separatory funnel and wash with 10 ml of cold water. Separate the organic layer and dry it over a few granules of anhydrous calcium chloride while you prepare for Part II. Transfer the residual liquid that remains in the distillation flask to a 25-ml Erlenmeyer, stopper, and label it "Crude Pinacolone". Save for a subsequent experiment. *The diene should be used during the same laboratory period in which it is prepared.*

PART II

In a 50-ml round-bottomed flask dissolve 5 g (0.05 mole) of finely pulverized maleic anhydride in 15 ml of anhydrous tetrahydrofuran. Warm if necessary to effect solution. Cool, then add 5 ml (4.2 g., 0.05 mole) of 2,3-dimethyl-1,3-butadiene. Swirl to mix. If the entire yield of Part I is taken, the diene should first be weighed and 1.17 g of maleic anhydride dissolved for each gram of diene taken. Shake the flask to mix thoroughly and note that the mixture becomes quite warm. Attach a reflux condenser to the flask and reflux for 30 minutes. Cool the reaction mixture in an ice-water bath, agitate, and scratch the sides of the flask with a stirring rod to initiate crystallization. The formation of crystals is sometimes a little slow. Collect your product on a Hirsch funnel, wash with 10 ml of cold water, and dry. Yield 3 g.; mp 78°C.

FIGURE 34.2 (a) Small-scale fractional distillation assembly showing Claisen adapter as a packed column; (b) glass-jointed drying tube as a packed column.

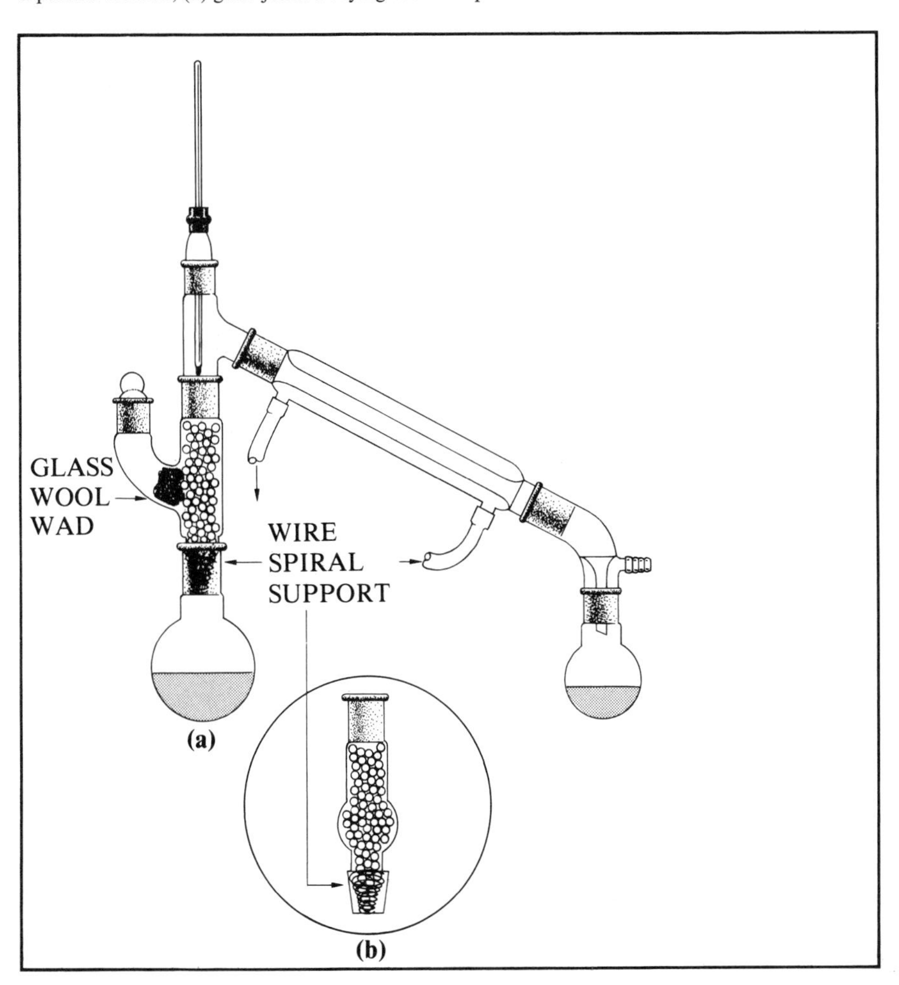

TEXT REFERENCE

Report: 34

Chapter Pages

Section Desk NAME

PREPARATION OF *cis*-4,5-DIMETHYL-1,2,3,6-TETRAHYDROPHTHALIC ANHYDRIDE

Reaction equations

PART I

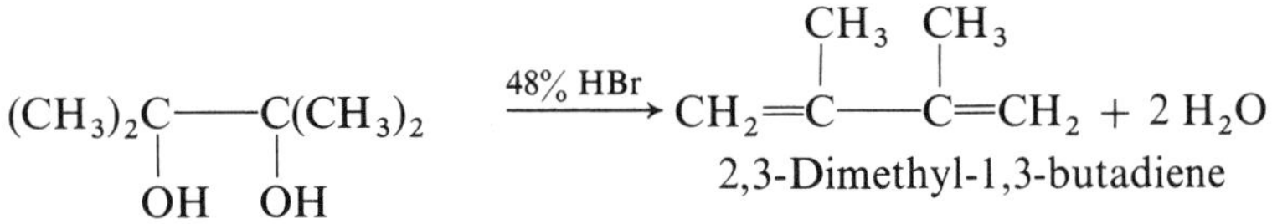

$$(CH_3)_2C(OH)\text{—}C(OH)(CH_3)_2 \xrightarrow{48\%\ HBr} CH_2{=}C(CH_3)\text{—}C(CH_3){=}CH_2 + 2\,H_2O$$

Pinacol → 2,3-Dimethyl-1,3-butadiene

PART II

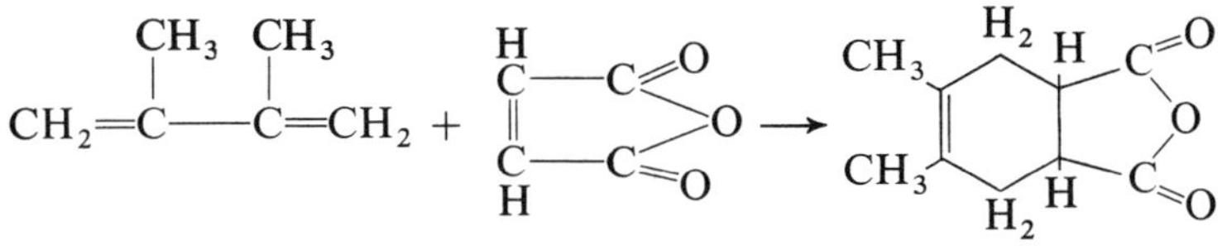

$$CH_2{=}C(CH_3)\text{—}C(CH_3){=}CH_2 + \text{Maleic anhydride} \rightarrow$$

Maleic anhydride

cis-4,5-Dimethyl-1,2,3,6-tetrahydrophthalic anhydride

	I	II	
Quantities	11.8 g	5 ml (4.2 g)	________ g
Mol. Wt.	________	________	________
Moles	________	________	________

Theoretical yield ________ g

Actual yield ________ g

Percentage yield ________

mp ________ °C

Questions and Exercises

1. How could you prepare benzoic acid using a Diels-Alder reaction as the first step in a three-part sequence?
2. Write structures for the adducts that result when the following compounds react with 2,3-dimethyl-1,3-butadiene.
   (a) acrylonitrile
   (b) acrolein
   (c) crotonaldehyde
   (d) *p*-benzoquinone
3. It was suggested in the procedure that 2,3-dimethyl-1,3-butadiene be used within the same laboratory period during which it is made. Why?

Experiment 35

Time: 3–4 hours

The Preparation of 2,2-Dimethylpropanoic Acid (Pivalic Acid)

One of the more interesting reactions exhibited by vicinal glycols is the facility with which they undergo intramolecular acid-catalyzed rearrangements. The mechanism of the rearrangement is fairly well established as one initiated by a protonation of one of the hydroxyl groups, followed by a dehydration and formation of a carbocation. Migration of a neighboring group *with its pair of electrons* to the electron-deficient carbon then results in a rearrangement of the carbon skeleton.

$$\underset{\text{Pinacol}}{CH_3-\overset{OH}{\underset{CH_3}{C}}-\overset{OH}{\underset{CH_3}{C}}-CH_3} \xrightarrow{H^+} CH_3-\overset{OH}{\underset{CH_3}{C}}-\overset{\overset{+}{O}H_2}{\underset{CH_3}{C}}-CH_3$$

$$\xrightarrow{-H_2O} CH_3-\overset{OH}{\underset{CH_3}{C}}-\overset{+}{\underset{CH_3}{C}}-CH_3$$

$$CH_3-\overset{OH}{\underset{CH_3}{C}}-\overset{+}{\underset{CH_3}{C}}-CH_3 \longrightarrow CH_3-\overset{\overset{+}{O}H}{\overset{\|}{C}}-\overset{CH_3}{\underset{CH_3}{C}}-CH_3$$

$$\xrightarrow{-H^+} \underset{\text{Pinacolone}}{CH_3-\overset{O}{\overset{\|}{C}}-\overset{CH_3}{\underset{CH_3}{C}}-CH_3}$$

The reaction is not limited to compounds that bear only hydroxyl groups on adjacent carbons but also occurs when amino and alkoxy groups appear on neighboring carbon atoms. All rearrangements of this type are referred to as pinacolic rearrangements. Migrating groups include, in addition to alkyl groups, aryl and hydrogen. From a number of studies of the reaction, the migratory aptitude of groups follows the order $Ar:^- > R:^- > H:^-$.

In the present experiment we will employ the pinacol–pinacolone rearrangement to prepare our starting material for synthesizing pivalic acid.

PART I THE PREPARATION OF PINACOLONE

In a 100-ml round-bottomed flask equipped with an adapter and condenser set for distillation place 65 ml of 6 *N* sulfuric acid and 11.3 g (0.05 mole) of previously prepared pinacol hydrate. Heat the mixture with a Bunsen burner and collect the distillate in a graduated cylinder. Continue the distillation until the upper organic layer in the graduated cylinder reaches a maximum value and only one phase remains in the distilling flask. The collection of 25 ml of distillate should be adequate. Transfer the distillate to a separatory funnel, separate, and discard the lower aqueous layer. Store the pinacolone in a small Erlenmeyer flask over a few granules of anhydrous calcium chloride, stopper, and save for Part II (Note 1).

PART II

In a 125-ml Erlenmeyer flask dissolve 4 g of sodium hydroxide in 35 ml of water and cool the solution in an ice-water bath. To the cold sodium hydroxide solution add 9.6 g (3.2 ml, 0.06 mole) of bromine from the bromine buret located in the hood. Swirl the flask until the bromine dissolves and the color of the solution is a light yellow. After the bromine has dissolved add 2.0 g (2.2 ml, 0.02 mole) of pinacolone. Stopper the flask and shake vigorously and intermittently for 30 minutes. Release the stopper slightly to vent any pressure after each shaking. The flask will become warm and the yellow color of the reaction mixture will fade completely. Transfer the mixture to a 100-ml round-bottomed flask and, using a Bunsen burner, remove the bromoform by steam distillation. Collect about 15 ml of distillate. Cool the distillation flask by immersing it in ice water, then transfer the cold contents to a separatory funnel. Extract the aqueous solution once with 15 ml of ether. **(EXTINGUISH ALL FLAMES!)** Discard the ether layer and collect the aqueous layer in a small Erlenmeyer flask. Set the flask in a pan of hot water and swirl periodically to drive out any dissolved ether. When the odor of ether can no longer be detected, set the flask in an ice bath and acidify the solution by the dropwise addition of concentrated sulfuric acid. If the solution turns yellow upon addition of acid, decolorize with a few drops of saturated sodium bisulfite solution. Continue to cool until crystallization is complete. Pivalic acid crystals will appear as prismatic needles. Collect on a Hirsch funnel. Yield 1–1.2 g; mp 34°C.

Note 1. A small amount of 2,3-dimethyl-1,3-butadiene (bp 76°C) will also appear in the distillate with pinacolone and may be removed by fractional distillation on a micro scale. A product distilling between 103–110°C should be collected as pinacolone (bp 106°C). However, the presence of the diene in pinacolone does not interfere with the haloform reaction that yields pivalic acid because any neutral organic compounds formed by bromination are removed in the extraction step before the acid is precipitated.

TEXT REFERENCE

Report: 35

Chapter Pages

Section Desk NAME

THE PREPARATION OF 2,2-DIMETHYLPROPANOIC ACID (PIVALIC ACID)

Reaction equations

$$\begin{matrix}(CH_3)_2C\text{—}OH \\ | \\ (CH_3)_2C\text{—}OH\end{matrix} \cdot 6\,H_2O \xrightarrow{H_3O^+} (CH_3)_3C\text{—}\overset{\overset{\displaystyle O}{\|}}{C}\text{—}CH_3 + 7\,H_2O$$

Pinacol hexahydrate — Pinacolone

$$(CH_3)_3C\text{—}\overset{\overset{\displaystyle O}{\|}}{C}\text{—}CH_3 + 3\,Br_2 + 4\,NaOH \rightarrow (CH_3)_3C\text{—}COO^-Na^+ + 3\,NaBr + 3\,H_2O$$

$$(CH_3)_3C\text{—}COO^-Na^+ + H_2SO_4 \rightarrow (CH_3)_3C\text{—}COOH + NaHSO_4$$

Pivalic acid

	Pinacolone	Pivalic acid
Quantities	2 g	________
Mol. Wt.	________	________
Moles	________	________

Theoretical yield ________ g

Actual yield ________ g

Percentage yield ________

mp ________°C

Questions and Exercises

1. Devise a synthesis for pinacolone that involves acetone but not pinacol.
2. Pinacolone will give a positive haloform reaction to yield chloroform or bromoform but fails to give a positive iodoform reaction. Why?
3. What property does pivalic acid have in common with other aliphatic acids of four or more carbons?
4. Devise a synthesis of pivalic acid that would not require the use of the haloform reaction in any step.
5. If you attempted a nitrile synthesis followed by hydrolysis in your solution to Question 4, your yield of pivalic acid would be very small indeed. What would be your principal product? Why?

Appendix

Suggested Equipment List

1 19/22 (Ŝ) 14-piece Glassware kit (see illustration, inside back cover)
1 Beaker, 400-ml
2 Beakers, 250-ml
1 Beaker, 150-ml
1 Beaker, 50-ml
2 Brushes (1 large, 1 small)
1 Burner, Bunsen with tubing
**1 Buret, 50-ml
4 Clamps, apparatus (with holders)
1 Screw clamp
2 Cylinders, graduated, 10-ml, and 100-ml
1 Dish, evaporating, 8 cm
**1 Flask, Erlenmeyer (wide necked) 500-ml
5 Flasks, Erlenmeyer; one 250-ml, two 125-ml, and two 50-ml
1 Flask, filtering, 125-ml
1 Funnel, Büchner (67 mm)
2 Funnels, glass 70-mm (short stem); 70-mm (long stem)
1 Funnel, Hirsch (42-mm)
Litmus, red (1 vial)
Litmus, blue (1 vial)
5 Medicine droppers
*2 Rings, iron; 3-in. and 4-in.
*2 Ring stands
*1 Steam bath, 6-in., 5-ring
2 Stirring rods, 5-mm × 200-mm
10 Test tubes, 150-mm, Pyrex
1 Test tube holder
1 Test tube rack
1 Thermometer, −5° to +260°C
**1 Thiele tube
1 Tube, drying
1 Watch glass, 75-mm
1 Wing top for burner
1 Wire gauze, asbestos center

* Commonly stored in the laboratory for general use.
** May be obtained from storeroom when needed.

PERSONAL SUPPLIES

1 Apron, laboratory, plastic or rubberized
1 Bottle or package of detergent or cleansing powder
1 Glass cutter or triangular file
Safety matches, 1 box
Safety glasses, goggles, or face shield
2 Spatulas, stainless
2 Towels, 18 × 24 inches

SUPPLEMENTARY EQUIPMENT FOR GENERAL USE

"Boileezers" or clay chips
Filter paper, $5\frac{1}{2}$-cm
First Aid Kit
Phenolphthalein indicator solution
Stopcock grease

Preparation of Special Reagents

BENEDICT'S SOLUTION

Dissolve 173 g of sodium citrate and 100 g of anhydrous sodium carbonate in 600 ml of water with heating. Dilute the volume to 850 ml and filter. To the filtered solution slowly add a second solution prepared from 17.3 g of copper sulfate ($CuSO_4 \cdot 5\ H_2O$) in 100 ml of water. Dilute the combined solutions to 1 liter.

FEHLING'S SOLUTION

Fehling's reagent consists of two solutions: Solution A is made by dissolving 69.69 g of copper sulfate ($CuSO_4 \cdot 5\ H_2O$) in distilled water and diluting to 1 liter; solution B is made by dissolving 346 g of sodium potassium tartrate and 250 g of sodium hydroxide in distilled water and diluting to 1 liter. Equal volumes of solutions A and B are mixed just prior to making a test.

IODINE-POTASSIUM IODIDE SOLUTION

Dissolve 100 g of iodine and 200 g of potassium iodide in 800 ml of distilled water.

LUCAS REAGENT

Dissolve 136 g (1.0 mole) of anhydrous zinc chloride in 105 g of concentrated hydrochloric acid with cooling.

MILLON'S REAGENT

Dissolve 50 g of mercury in 100 g of nitric acid (sp. gr. 1.40) and warm until solution is complete. Dilute with 200 ml of water.

NINHYDRIN REAGENT

Dissolve 0.4 g of ninhydrin and 1.5 ml pyridine in 100 ml of 95% ethanol.

SCHIFF'S REAGENT

Dissolve 1.0 g of fuchsin (*p*-rosaniline hydrochloride) in 700 ml of warm distilled water. Filter if necessary. Cool the solution and pass sulfur dioxide into it until colorless. Dilute to 1 liter.

Chemicals and Reagents Required (By Experiments)

Experiment		Approximate Quantities for 10 Students[1]
1.	**MELTING POINT DETERMINATION**	
	Mineral oil (high boiling)	1,000 ml
	17 Common organic compounds listed in Table 1.1	5 g each
	Capillary melting point tubes	60
2.	**CRYSTALLIZATION**	
	Acetanilide (Impure)[2]	20 g
	Decolorizing charcoal (Norite)	5 g
3.	**BOILING POINTS — DISTILLATION**	
	2-Propanol	250 ml
	Ethyl alcohol (95%)	300 ml

[1]Ordinary desk reagents are not included. The following reagents are kept in the laboratory on the side shelf or on the reagent racks over the students' desks: Sulfuric acid, concd. and dilute; Hydrochloric acid, concd. and dilute; Nitric acid, concd. and dilute; Ammonium hydroxide, concd.; Acetic acid, glacial.

[2]Impure acetanilide for this experiment may be prepared by mixing intimately 100 g acetanilide, 5 g brown sugar, and 5 g wood flour or fine sawdust.

4. FRACTIONAL DISTILLATION

Acetone	150 ml

5. EXTRACTION — USE OF THE SEPARATORY FUNNEL

Mandelic acid	80 g
Phenolphthalein solution	10 ml
0.3 *N* Sodium hydroxide	1,500 ml
Ethyl ether	600 ml
Benzoic acid, *trans*-cinnamic acid, biphenyl, naphthalene, *p*-toluidine, *p*-chloroaniline	5 g each
Magnesium sulfate (anhydrous)	10 g
Sodium hydroxide (10%)	500 ml
Ether	300 ml

6. CHROMATOGRAPHY

Part I

Glass wool	5 g
Glass tubing, 15-mm	9 ft
Aluminum oxide (chromatographic grade, Aluminum Company of America or Merck)	200 g
Hyflo Super-cel or Celite (Johns-Manville)	50 g
Ethyl alcohol (95%)	500 ml
n-Propyl alcohol	variable
Acetone	variable
Sudan III	1 g
Methylene blue	1 g

Part II

Whatman #1 filter paper for chromatography 10 × 16 cm	10 sheets
Micropipettes or melting point capillaries	30
Four assorted food colors (from local food market)	1 carton
Ethyl alcohol (95%)	100 ml
Isopropyl alcohol	100 ml
Aluminum foil	1 roll

Part III

Either:	
Microscope slides (75 × 25 mm)	60
Silica gel (Swiss type D5 Kieselgel, Arthur H. Thomas Co.) or (SilicAR TLC-7GF, Mallinckrodt Chemical Works.)	20 g
Or:	
Eastman Kodak Chromatography Film (K301R2) 200 × 200 mm	2 sheets

n-Propyl alcohol	25 ml
Acetone	5 ml
Plastic electrician's tape (¾ inch) or special tape for TLC sold by Mallinckrodt Chemical Works	1 roll
Micropipettes or melting point capillaries	30
Toluene	60 ml
Ether	5 ml
Iodine	1 g

Dye mixture:

Butter yellow Rhodamine B Methylene Blue	3% w/v of each in 95% ethyl alcohol	1 ml

Phenol, 6% w/v in ethyl alcohol	1 ml
Catechol, 6% w/v in ethyl alcohol	1 ml
Resorcinol, 6% w/v in ethyl alcohol	1 ml
Pyrogallol, 6% w/v in ethyl alcohol	1 ml

PART IV

Cyclohexane	variable
2-Propanol	variable

7. REACTIONS OF THE HYDROCARBONS

"Skellysolve C" or Petroleum ether (boiling range 70–90°)	100 ml
Potassium permanganate solution (0.5–1.0%)	50 ml
Bromine in carbon tetrachloride solution	50 ml
Cyclohexene	100 ml
Toluene	50 ml
Sodium chloride (saturated)	100 ml
Bromine	5 ml
Calcium carbide	50 g
Silver nitrate (2.0%)	50 ml
Anhydrous toluene	50 ml
Sodium metal	1.0 g
Potassium permanganate (0.3–1.0%)	50 ml
Bromine in methylene chloride	20 ml

8. STEAM DISTILLATION: THE ISOLATION OF LIMONENE FROM ORANGE PEEL

Oranges (large navel)	10
Bromine in methylene chloride solution	50 ml
Absolute alcohol	250 ml

9. A STUDY OF REACTION RATES. THE HYDROLYSIS OF tert-BUTYL HALIDES

Acetone (reagent grade)	1,000 ml
tert-Butyl chloride	50 ml
tert-Butyl bromide	50 ml
0.1 *M* Sodium hydroxide solution	100 ml
Bromophenol blue indicator	2 ml

10. ALCOHOLS AND PHENOLS

(I) PROPERTIES AND REACTIONS OF ALCOHOLS

Methanol	25 ml
Ethyl alcohol (95%)	25 ml
Isopropyl alcohol	45 ml
n-Propyl alcohol	25 ml
n-Butyl alcohol	10 ml
sec-Butyl alcohol	10 ml
tert-Butyl alcohol	30 ml
Cyclohexanol	10 ml
Benzyl alcohol	10 ml
Sodium hydroxide (5%)	50 ml
Phosphoric acid (85%)	200 ml
Sodium metal	2 g
Lucas reagent	20 ml
Potassium dichromate	50 g
Acetone	5 ml
Iodine in potassium iodide solution	10 ml
Sodium hydroxide (6 *N*)	25 ml
Samples of alcohols listed in Table 13.1 to be used as unknowns	5 ml each

(II) PROPERTIES AND REACTIONS OF PHENOLS

Phenol	2 g
p-Chlorophenol, 2-naphthol, resorcinol, and salicylic acid	1 g each
Ethyl alcohol	5 ml
Ethyl acetoacetate	5 ml
Sodium hydroxide (5%)	50 ml
Bromine water (saturated)	25 ml
Ferric chloride (1.0%)	5 ml

11. REACTIONS OF ALDEHYDES AND KETONES

Item	Amount
Copper wire (#14)	8 ft
Methyl alcohol	20 ml
Ethyl alcohol (95%)	100 ml
Paraldehyde	100 ml
Acetone	25 ml
Benzaldehyde	5 ml
Cyclopentanone	5 ml
Diethyl ketone	5 ml
Schiff's reagent	50 ml
Fehling's or Benedict's solution	50 ml
Silver nitrate (5%)	20 ml
Formalin (35–40% formaldehyde)	10 ml
Sodium bisulfite (saturated)	50 ml
Benzophenone	45 g
Sodium borohydride	10 g
Methanol	300 ml
Ethyl alcohol (95%)	variable

12. IDENTIFICATION OF AN UNKNOWN CARBONYL COMPOUND

Item	Amount
Phenylhydrazine hydrochloride	20 g
2,4-Dinitrophenylhydrazine	5 g
Sodium acetate	60 g
Benzaldehyde	5 ml
Ethyl alcohol (95%)	50 ml
Methyl ethyl ketone	10 ml
Semicarbazide hydrochloride	20 g
7 carbonyl compounds listed in Table 12.1 to be used as unknowns	5 ml samples

13. REACTIONS OF THE AMINES

Item	Amount
Aniline	10 ml
Benzylamine	15 ml
Diethylamine	5 ml
Pyridine	5 ml
Ether	100 ml
Sodium Chloride	variable
Silver nitrate (2%)	20 ml
Benzoyl chloride	5 ml
Sodium hydroxide (10%)	20 ml
Ethyl alcohol	150 ml
Phenyl isothiocyanate	5 ml
n-Hexane or Skellysolve B	50 ml

14. IDENTIFICATION OF UNKNOWN ORGANIC COMPOUNDS: IR AND NMR SPECTROSCOPY

24 common organic compounds listed in Table 14.1. Some of these may be student preparations from previous experiments .... 5-ml or 5-g samples

15. FATS AND OILS; SOAPS AND DETERGENTS

Cottonseed oil	5 ml
Crisco	60 g
Methylene chloride	100 ml
Bromine in methylene chloride (5% solution)	50 ml
Linseed oil (boiled)	5 g
Mazola (any cooking oil may be substituted)	5 g
Sodium hydroxide	25 g
Ethyl alcohol (95%)	150 ml
Sodium chloride	300 g
Ivory Flakes or Dreft	5 g
Calcium chloride (0.1%)	10 ml
Magnesium chloride (0.1%)	10 ml
Ferric chloride (0.1%)	10 ml
Mineral oil	5 ml
Congo red indicator paper	1 vial
Stearic acid	2.5 g

16. CARBOHYDRATES

Glucose	10 g
Fructose	10 g
Sucrose	10 g
Maltose	10 g
Lactose	10 g
Starch	10 g
Molisch reagent	variable
Fehling's or Benedict's solution	200 ml
Phenylhydrazine hydrochloride	20 g
Sodium acetate	30 g
Acetone	150 ml
Ethyl alcohol (95%)	150 ml
Acetic anhydride	70 ml
Ether	100 ml
Sodium hydroxide (10%)	variable
Absorbent cotton	10 g
Dibutyl phthalate	2 ml

17. PROTEINS

Item	Quantity
Casein	10 g
Egg albumin	10 g
Copper sulfate (2%)	5 ml
Glycine	1 g
Sodium nitrite (5%)	50 ml
Soda lime	20 g
Eggs	2
Urea	20 g
Copper sulfate (10%)	25 ml
Ferric chloride (10%)	25 ml
Lead acetate (10%)	25 ml
Mercuric chloride (10%)	25 ml
Lead acetate test paper	1 vial
Whatman #1 filter paper (10 × 16 cm)	10 sheets
Samples of various amino acids; including aspartic acid, phenylalanine, and tyrosine	1 g each
Phenol (80% solution)	150 ml
Acetone (in polyethylene squeeze bottle)	125 ml
Either:	
Atomizer (or "Windex" glass cleaner spray bottle)	1
Ethyl alcohol (95%)	100 ml
Ninhydrin	2 g
Or:	
"Ninspray" Reagent (Nutritional Biochemical Corp., Cleveland, Ohio 44128)	1 can
Throw-away plastic gloves ("Handgards," Plasticsmith, Inc., Pittsburg, Calif.)	20

18. BIOSYNTHESIS OF ETHANOL AND ACETIC ACID

Item	Quantity
Freshly prepared apple cider	1 gallon
or	
Frozen apple juice concentrate	1 12-ounce can
Saturated $Ca(OH)_2$ solution	1 liter
(Reagents listed under Experiment 10 for carrying out the oxidation and iodoform tests)	
Sodium hydroxide solution (standardized)	1 liter

19. CYCLOHEXENE

Item	Quantity
Cyclohexanol	200 g
Phosphoric acid (85%)	50 ml
Calcium chloride (anhydrous)	30 g
Sodium chloride	variable

Sodium carbonate (10%)	variable
Potassium permangante (0.5%)	5 ml
Bromine in methylene chloride	5 ml

20. PREPARATION OF ALKYL HALIDES

(A) *tert*-BUTYL CHLORIDE

tert-Butyl alcohol	200 ml
Sodium carbonate (5%)	variable
Calcium chloride (anhydrous)	20 g

(B) *n*-BUTYL BROMIDE

Sodium bromide	310 g
n-Butyl alcohol	185 g
Sodium carbonate (10%)	250 ml
Calcium chloride (anhydrous)	50 g

(C) CLASSIFICATION TESTS FOR ALKYL HALIDES

Silver nitrate (2% in 95% ethanol)	60 ml
Sodium iodide (15% w/v in anhydrous acetone)	40 ml
n-Butyl bromide	2 ml
t-Butyl chloride	2 ml

21. THE GRIGNARD REACTION: THE PREPARATION OF 2-METHYL-2-HEXANOL

Magnesium metal	30 g
Ether (anhydrous)	1,000 ml
n-Butyl bromide	110 ml
Iodine	variable
Acetone (anhydrous)	100 ml
Ammonium chloride (saturated)	750 ml
Sodium bisulfite (saturated)	250 ml
Sodium chloride (saturated)	250 ml
Magnesium sulfate (anhydrous)	variable

22. CHROMIC ACID OXIDATION: PREPARATION OF ADIPIC ACID

Potassium dichromate ($K_2Cr_2O_7$)	90 g
Cyclohexene (if not prepared in Experiment 19)	40 ml

23. ACID DERIVATIVES

(I) PREPARATION OF *trans*-CINNAMAMIDE

Cinnamic acid	30 g
Cinnamoyl chloride (if not prepared in expt)	30 g
Thionyl chloride	25 g
Methanol	100 ml

(II) PHENYLACETAMIDE

Phenylacetic acid	28 g
Phenylacetyl chloride (if not prepared in expt)	30 g
Thionyl chloride	25 g
Methanol	100 ml

(III) PREPARATION OF BENZOIC ANHYDRIDE

Benzoyl chloride	20 ml
Pyridine	80 ml
"Skellysolve C"	variable

24. ISOAMYL ACETATE

Isoamyl alcohol	220 ml
Magnesium sulfate (anhydrous)	30 g
Sodium bicarbonate (10%)	300 ml

25. ASPIRIN AND OIL OF WINTERGREEN

(I) ASPIRIN

Salicylic acid	20 g
Acetic anhydride	50 ml
Phosphoric acid (85%)	5 ml
Methanol	10 ml
Ferric chloride (1.0%)	5 ml

(II) METHYL SALICYLATE

Salicylic Acid	70 g
Methyl alcohol	300 ml
Sodium bicarbonate (5%)	500 ml
Calcium chloride	10 g

26. ACETANILIDE

Aniline	200 g
Decolorizing charcoal	20 g
Sodium acetate ($CH_3COONa \cdot 3\ H_2O$)	320 g
Acetic anhydride	250 ml

27. THE FRIEDEL-CRAFTS REACTION

(A) *p*-ANISYL BENZYL KETONE

Phenylacetic acid	14 g
Polyphosphoric acid	150 g
Anisole	12 g
Sodium bicarbonate (10%)	50 ml
Methanol	200 ml

(B) BENZYL PHENYL KETONE

Benzoic acid	24 g
Polyphosphoric acid	200 g
Anisole	16 g
Ether	25 ml
Sodium hydroxide (10%)	500 ml
Methanol	200 ml

28. **BENZANILIDE**

Benzophenone	19 g
Polyphosphoric acid	200 g
Hydroxylamine hydrochloride	21 g
Ethyl alcohol	200 ml

29. **POLYMERIZATION**

(A) POLYURETHANE FOAM

Castor oil	350 g
Glycerol	100 g
Stannous octoate	5 ml
Silicone oil (Dow-Corning 202)	10 ml
4-Methyl-*m*-phenylene diisocyanate (Toluene 2,4-diisocyanate)	300 g
Paper cups or waxed cartons	10

(B) NYLON 6-10

Sebacoyl chloride (5% w/v in carbon tetrachloride)	40 ml
Hexamethylene diamine (5% w/v in water)	40 ml

30. **THE COUPLING OF AROMATIC DIAZONIUM COMPOUNDS; DYES AND DYEING**

Sodium nitrite	10 g
β-Naphthol	10 g
Malachite green	1 g
Methyl violet	1 g
Tannic acid	2 g
Potassium antimony tartrate	2 g
Picric acid	1 g
Samples of cotton cloth or bandage gauze, 1 × 2 in.	50
Samples of wool and silk, 1 × 2 in.	10 each

31. **PREPARATION OF *tert*-BUTYLBENZOIC ACID**

tert-Butylbenzene	350 ml
Bromine	240 g
Upholsterer's tacks (no. 5)	50
Sodium carbonate (5%)	750 ml

Calcium chloride (anhydrous) .... 100 g
Magnesium metal .... 30 g
Ether (anhydrous) .... 1,000 ml
Iodine .... variable
Dry ice .... 400 g
Sodium hydroxide (5%) .... 1,500 ml
Sodium bisulfate (satuarated) .... 200 ml

32. **SULFANILIDE**

Chlorosulfonic acid .... 125 ml
Acetanilide .... 50 g
Vegetable charcoal .... 5 g

33. **PINACOL HYDRATE**

Tetrahydrofuran (reagent grade) .... 1,000 ml
Magnesium metal .... 120 g
Acetone (reagent grade) .... 1,000 ml
Sodium chloride (saturated) .... 2,000 ml
Petroleum ether (boiling range 50–70°) .... 2,000 ml

34. **THE DIELS-ALDER REACTION**

Pinacol hydrate .... 150 g (or student preparation)
Cyclohexane .... 750 ml
Tetrahydrofuran .... 150 ml
Maleic anhydride .... 50–75 g

35. **PREPARATION OF 2,2-DIMETHYLPROPANOIC ACID**

Pinacol hydrate .... 150 g (or student preparation)
Sodium hydroxide .... 40 g
Bromine .... 100 g
Ether .... 150 ml
Sodium bisulfite (saturated solution) .... 50 ml

Infrared and Nuclear Magnetic Resonance Spectra of Compounds Listed in Table 14.1

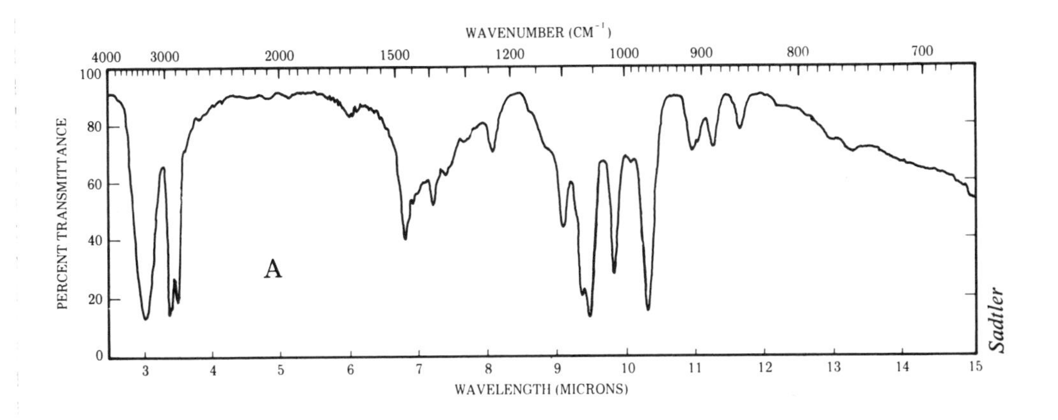

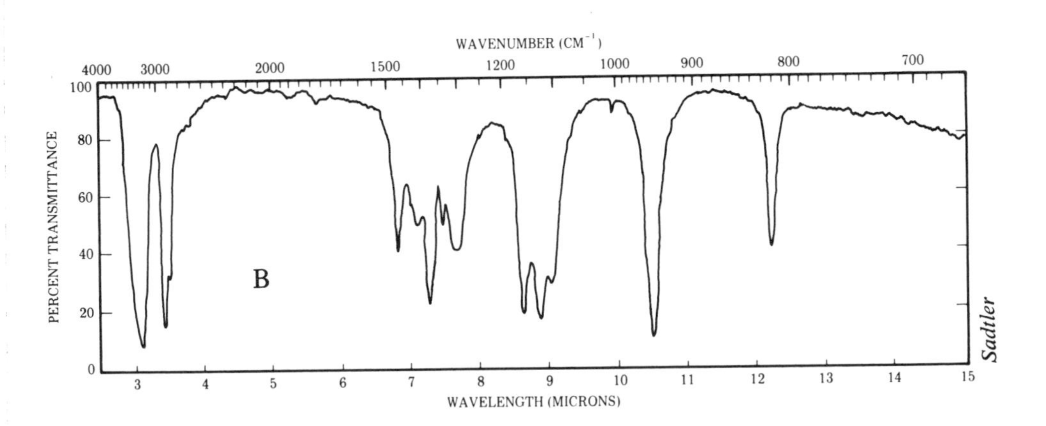

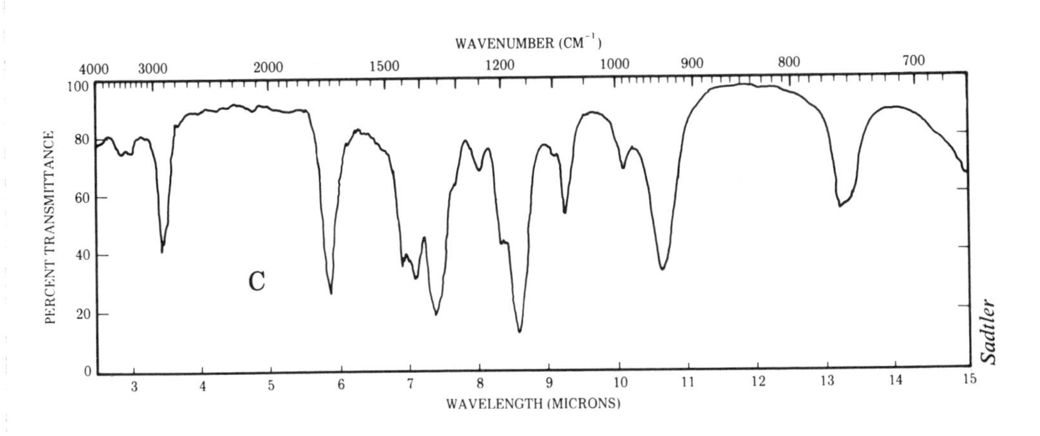

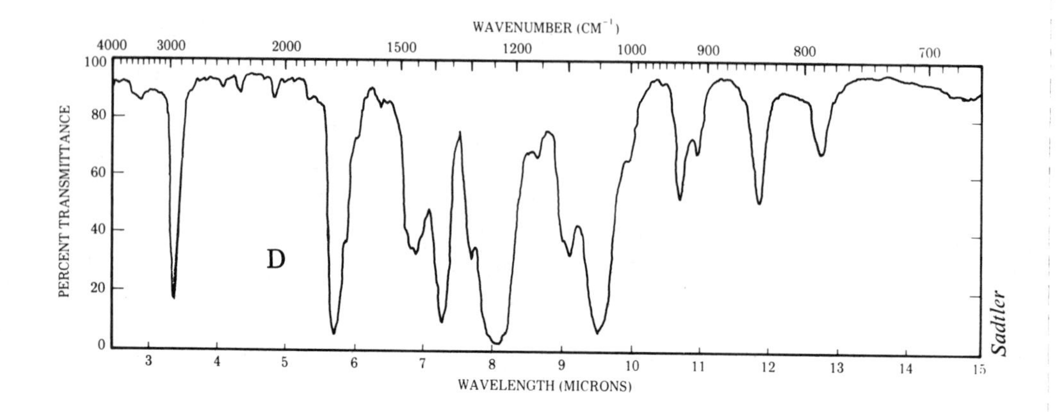
WAVENUMBER (CM⁻¹)
4000 3000 2000 1500 1200 1000 900 800 700
PERCENT TRANSMITTANCE
D
WAVELENGTH (MICRONS)
Sadtler

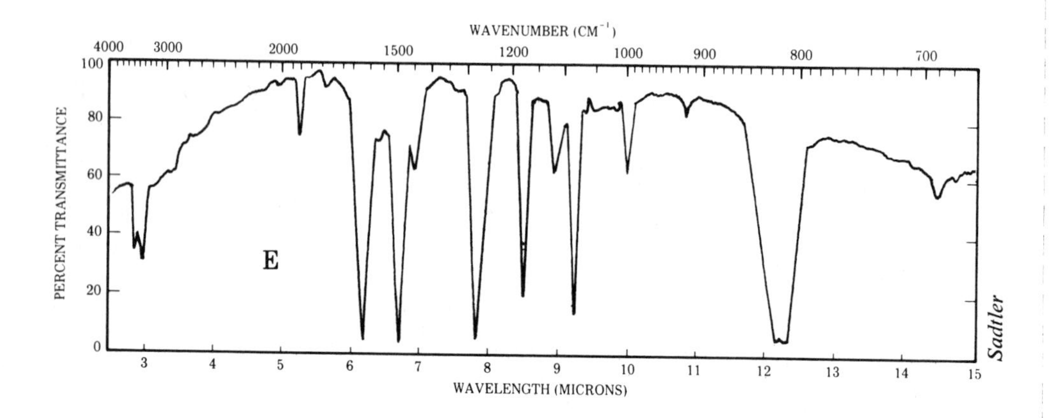
WAVENUMBER (CM⁻¹)
4000 3000 2000 1500 1200 1000 900 800 700
PERCENT TRANSMITTANCE
E
WAVELENGTH (MICRONS)
Sadtler

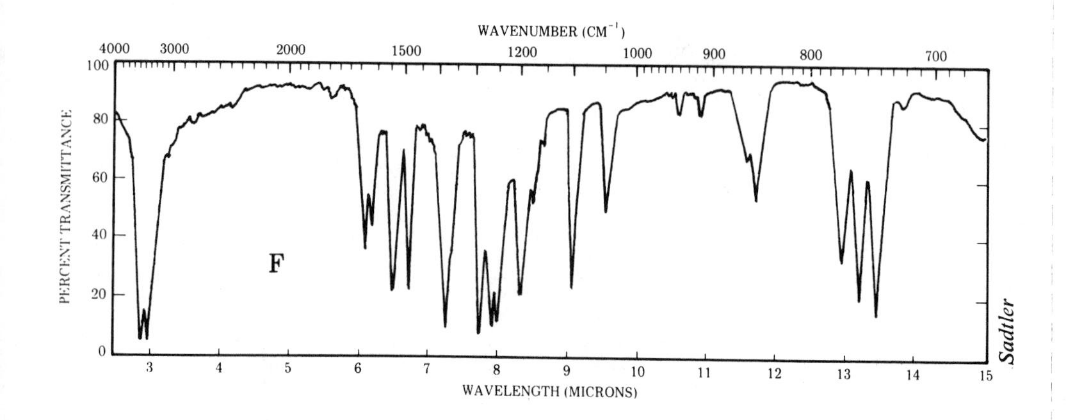
WAVENUMBER (CM⁻¹)
4000 3000 2000 1500 1200 1000 900 800 700
PERCENT TRANSMITTANCE
F
WAVELENGTH (MICRONS)
Sadtler

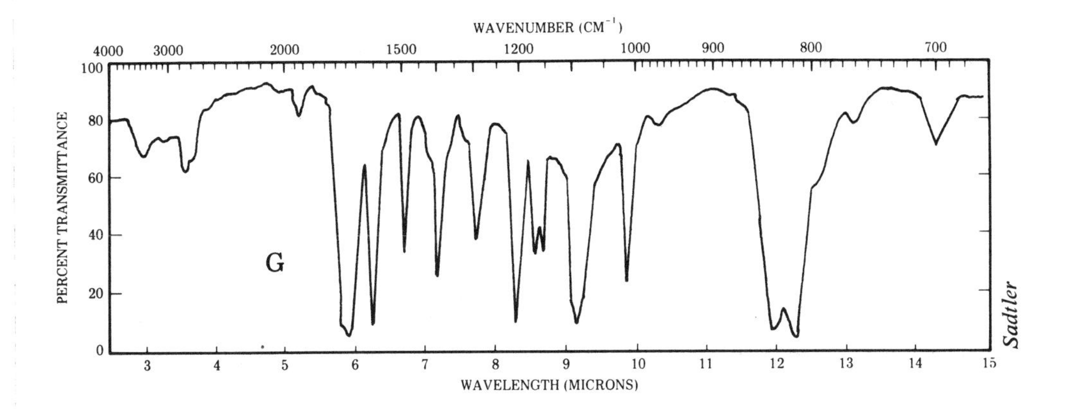
WAVENUMBER (CM^{-1})
4000 3000 2000 1500 1200 1000 900 800 700
PERCENT TRANSMITTANCE
100 80 60 40 20 0
G
3 4 5 6 7 8 9 10 11 12 13 14 15
WAVELENGTH (MICRONS)
Sadtler

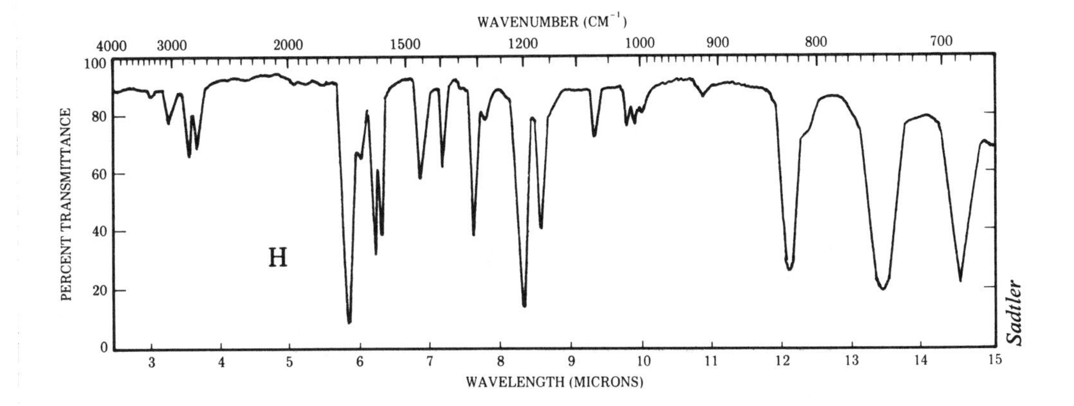
WAVENUMBER (CM^{-1})
4000 3000 2000 1500 1200 1000 900 800 700
PERCENT TRANSMITTANCE
100 80 60 40 20 0
H
3 4 5 6 7 8 9 10 11 12 13 14 15
WAVELENGTH (MICRONS)
Sadtler

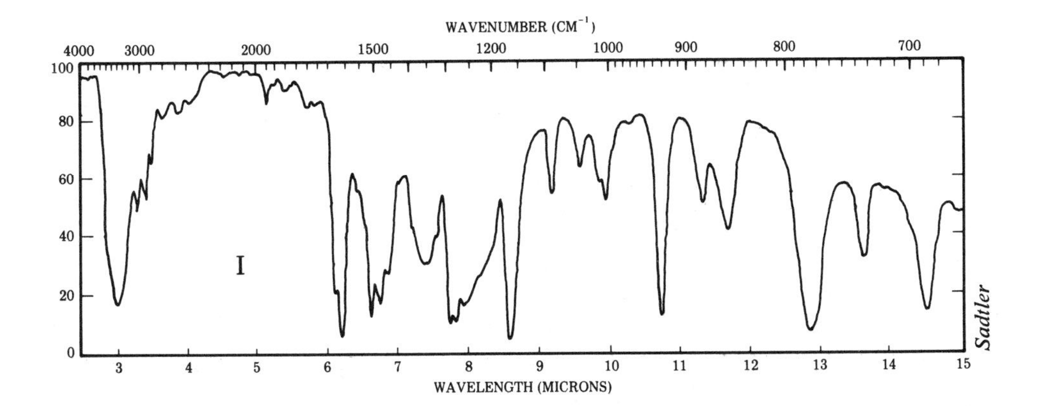
WAVENUMBER (CM^{-1})
4000 3000 2000 1500 1200 1000 900 800 700
100 80 60 40 20 0
I
3 4 5 6 7 8 9 10 11 12 13 14 15
WAVELENGTH (MICRONS)
Sadtler

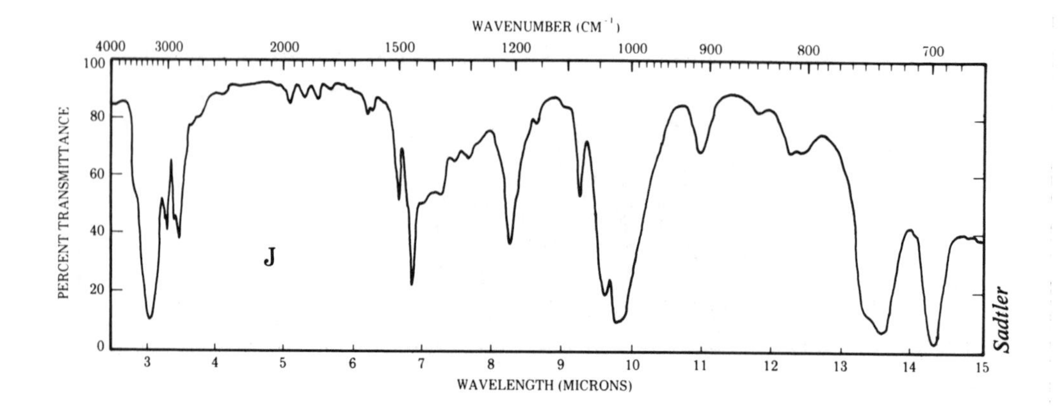
WAVENUMBER (CM⁻¹)
4000 3000 2000 1500 1200 1000 900 800 700
100 80 60 40 20 0
PERCENT TRANSMITTANCE
J
Sadtler
3 4 5 6 7 8 9 10 11 12 13 14 15
WAVELENGTH (MICRONS)

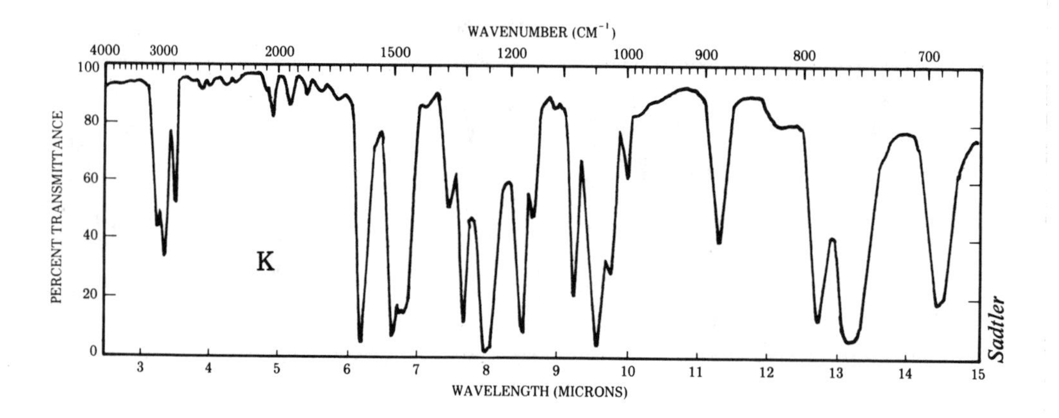
WAVENUMBER (CM⁻¹)
4000 3000 2000 1500 1200 1000 900 800 700
100 80 60 40 20 0
PERCENT TRANSMITTANCE
K
Sadtler
3 4 5 6 7 8 9 10 11 12 13 14 15
WAVELENGTH (MICRONS)

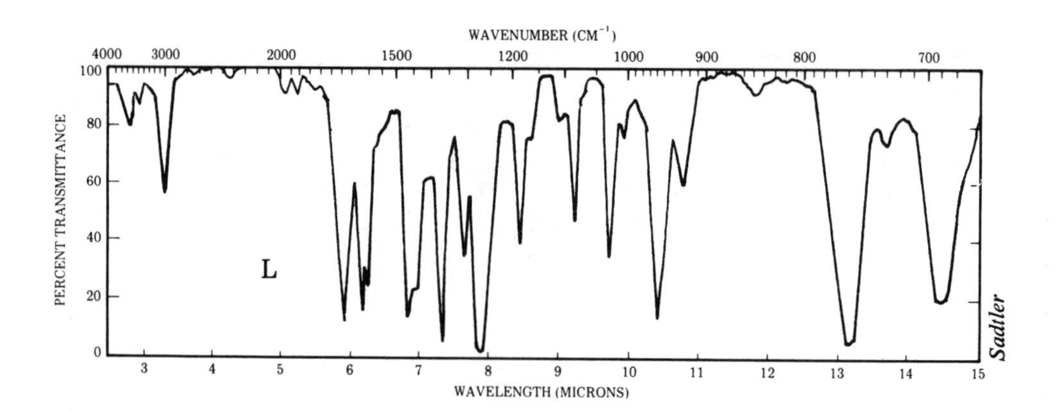
WAVENUMBER (CM⁻¹)
4000 3000 2000 1500 1200 1000 900 800 700
100 80 60 40 20 0
PERCENT TRANSMITTANCE
L
Sadtler
3 4 5 6 7 8 9 10 11 12 13 14 15
WAVELENGTH (MICRONS)

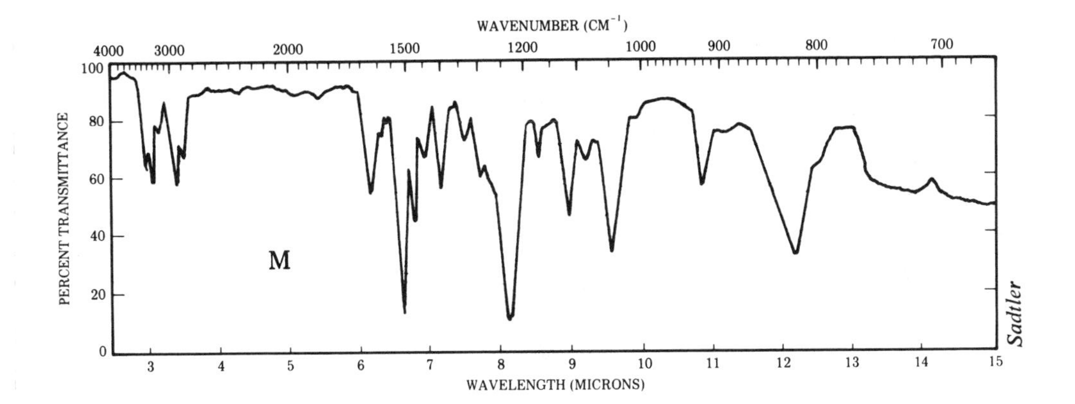
WAVENUMBER (CM⁻¹)
4000 3000 2000 1500 1200 1000 900 800 700
PERCENT TRANSMITTANCE
100 80 60 40 20 0
M
Sadtler
3 4 5 6 7 8 9 10 11 12 13 14 15
WAVELENGTH (MICRONS)

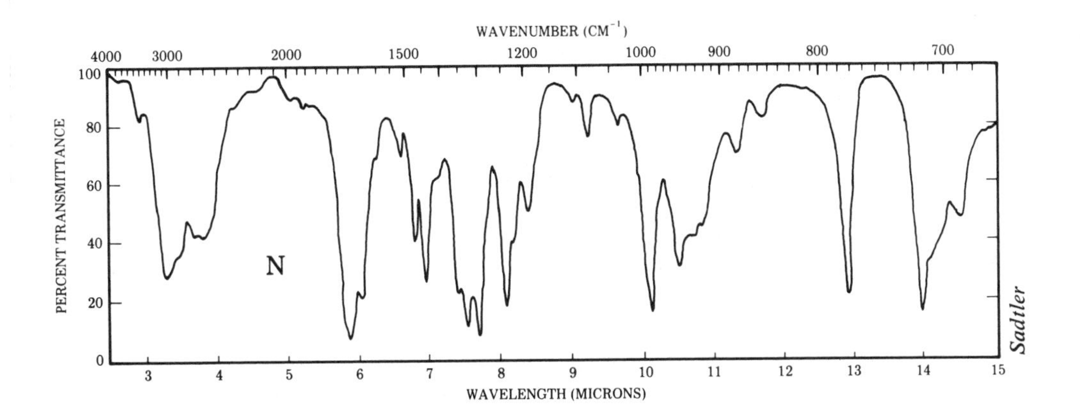
WAVENUMBER (CM⁻¹)
4000 3000 2000 1500 1200 1000 900 800 700
PERCENT TRANSMITTANCE
100 80 60 40 20 0
N
Sadtler
3 4 5 6 7 8 9 10 11 12 13 14 15
WAVELENGTH (MICRONS)

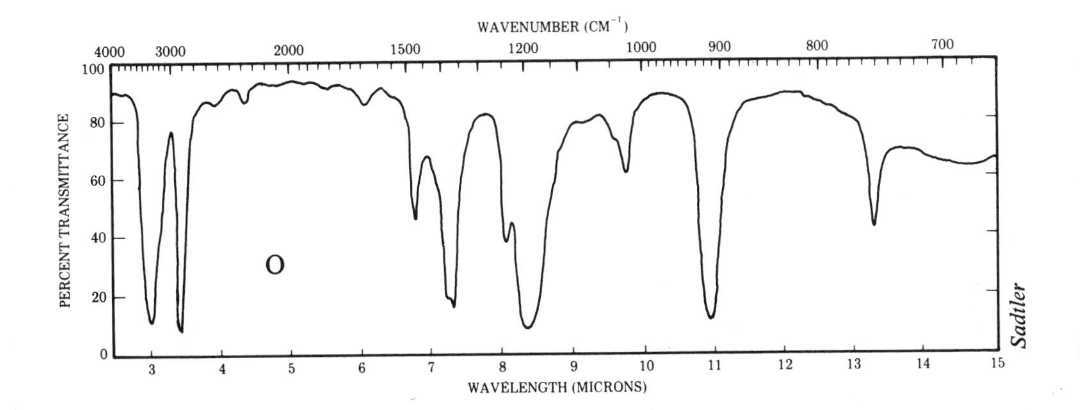
WAVENUMBER (CM⁻¹)
4000 3000 2000 1500 1200 1000 900 800 700
PERCENT TRANSMITTANCE
100 80 60 40 20 0
O
Sadtler
3 4 5 6 7 8 9 10 11 12 13 14 15
WAVELENGTH (MICRONS)

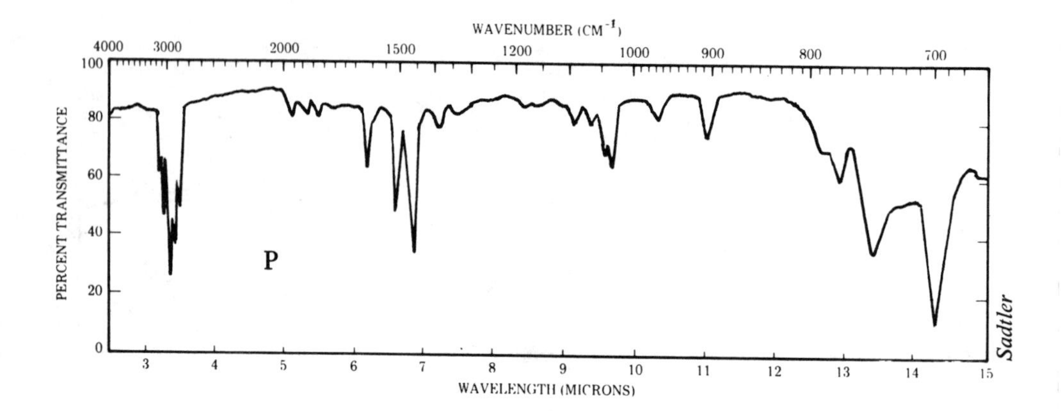
WAVENUMBER (CM⁻¹)
4000 3000 2000 1500 1200 1000 900 800 700
PERCENT TRANSMITTANCE
100 80 60 40 20 0
P
3 4 5 6 7 8 9 10 11 12 13 14 15
WAVELENGTH (MICRONS)
Sadtler

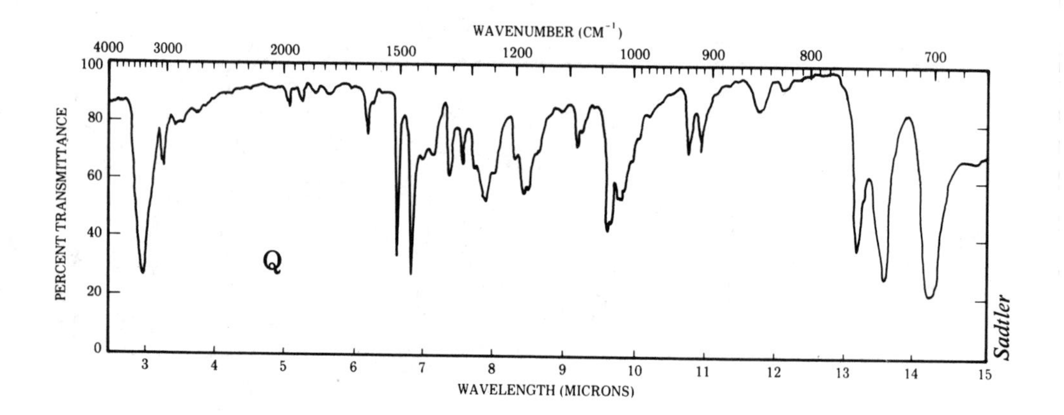
WAVENUMBER (CM⁻¹)
4000 3000 2000 1500 1200 1000 900 800 700
PERCENT TRANSMITTANCE
100 80 60 40 20 0
Q
3 4 5 6 7 8 9 10 11 12 13 14 15
WAVELENGTH (MICRONS)
Sadtler

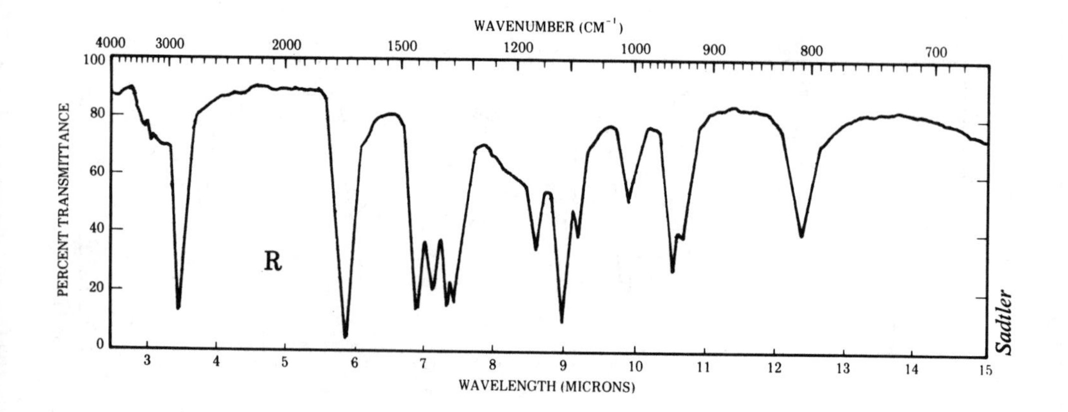
WAVENUMBER (CM⁻¹)
4000 3000 2000 1500 1200 1000 900 800 700
PERCENT TRANSMITTANCE
100 80 60 40 20 0
R
3 4 5 6 7 8 9 10 11 12 13 14 15
WAVELENGTH (MICRONS)
Sadtler

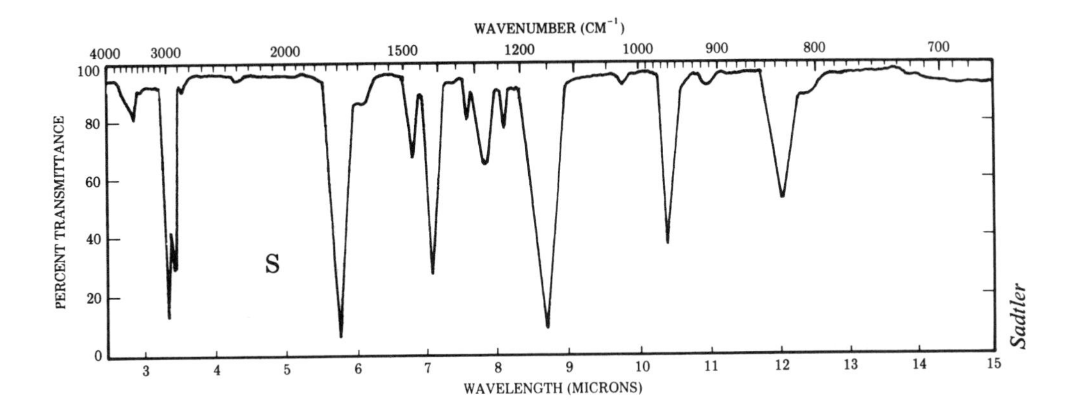
WAVENUMBER (CM⁻¹)
4000 3000 2000 1500 1200 1000 900 800 700
PERCENT TRANSMITTANCE
100 80 60 40 20 0
S
3 4 5 6 7 8 9 10 11 12 13 14 15
WAVELENGTH (MICRONS)
Sadtler

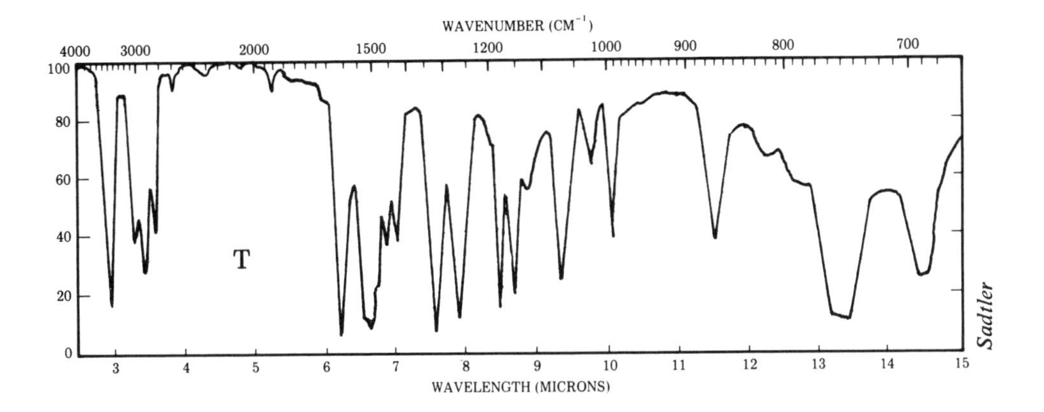
WAVENUMBER (CM⁻¹)
4000 3000 2000 1500 1200 1000 900 800 700
100 80 60 40 20 0
T
3 4 5 6 7 8 9 10 11 12 13 14 15
WAVELENGTH (MICRONS)
Sadtler

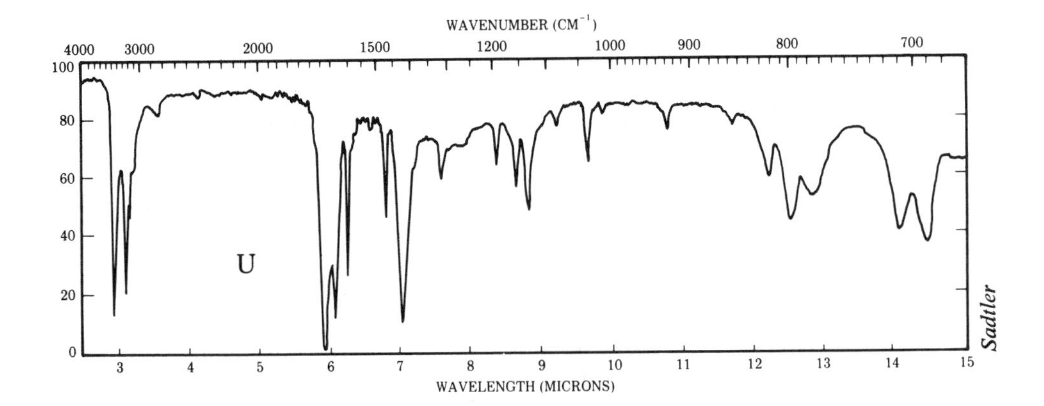
WAVENUMBER (CM⁻¹)
4000 3000 2000 1500 1200 1000 900 800 700
100 80 60 40 20 0
U
3 4 5 6 7 8 9 10 11 12 13 14 15
WAVELENGTH (MICRONS)
Sadtler

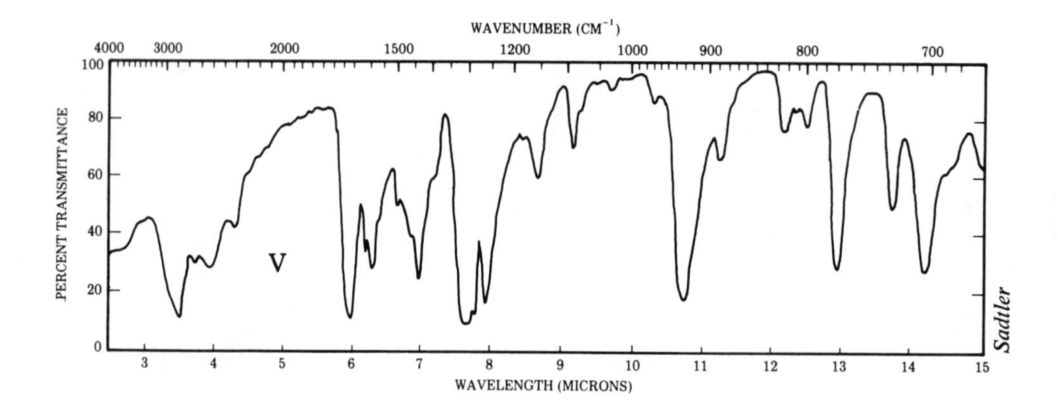
WAVENUMBER (CM⁻¹)
4000 3000 2000 1500 1200 1000 900 800 700
PERCENT TRANSMITTANCE
100 80 60 40 20 0
V
3 4 5 6 7 8 9 10 11 12 13 14 15
WAVELENGTH (MICRONS)
Sadtler

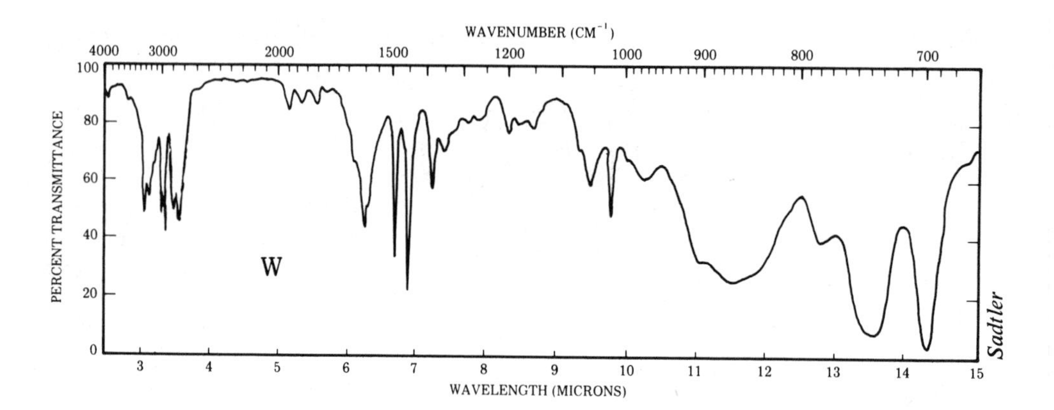
WAVENUMBER (CM⁻¹)
4000 3000 2000 1500 1200 1000 900 800 700
PERCENT TRANSMITTANCE
100 80 60 40 20 0
W
3 4 5 6 7 8 9 10 11 12 13 14 15
WAVELENGTH (MICRONS)
Sadtler

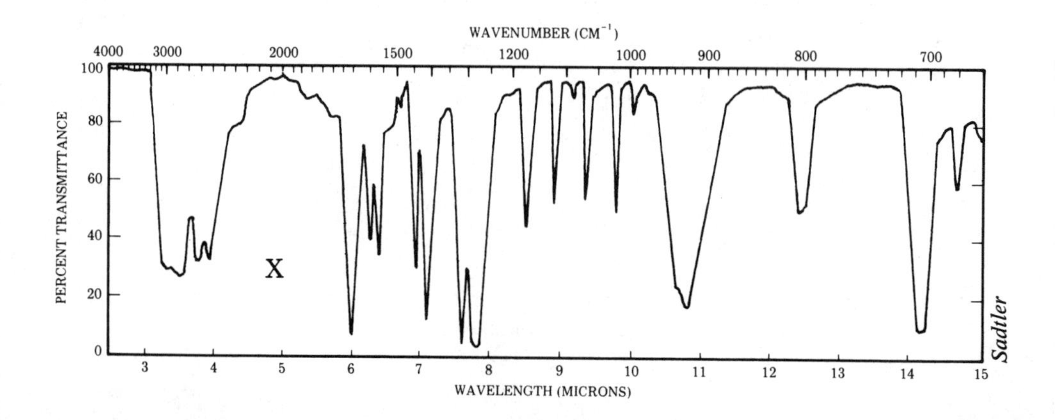
WAVENUMBER (CM⁻¹)
4000 3000 2000 1500 1200 1000 900 800 700
PERCENT TRANSMITTANCE
100 80 60 40 20 0
X
3 4 5 6 7 8 9 10 11 12 13 14 15
WAVELENGTH (MICRONS)
Sadtler

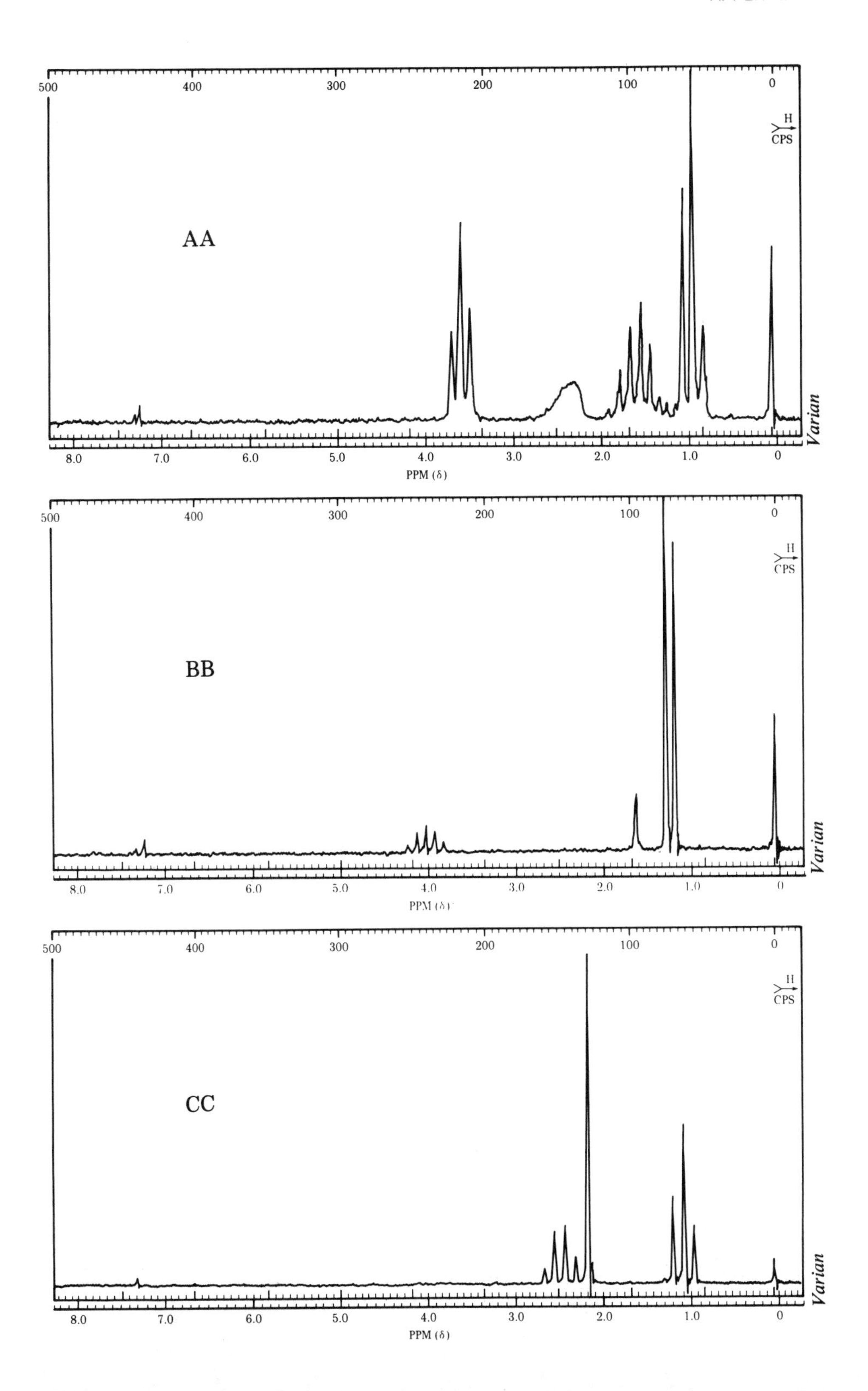
AA
BB
CC
PPM (δ)
CPS
Varian

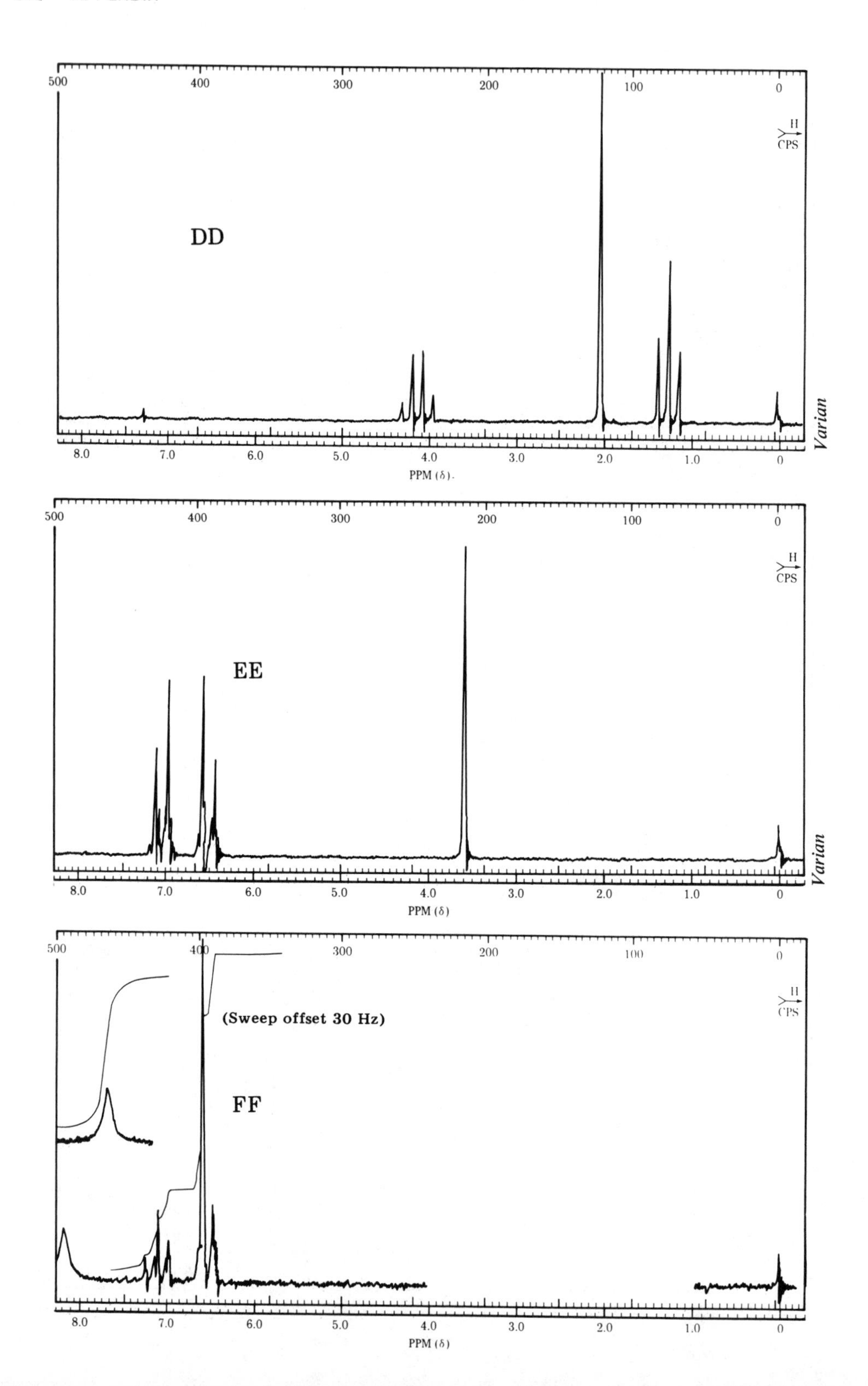

DD
EE
FF
(Sweep offset 30 Hz)
PPM (δ)
CPS
Varian

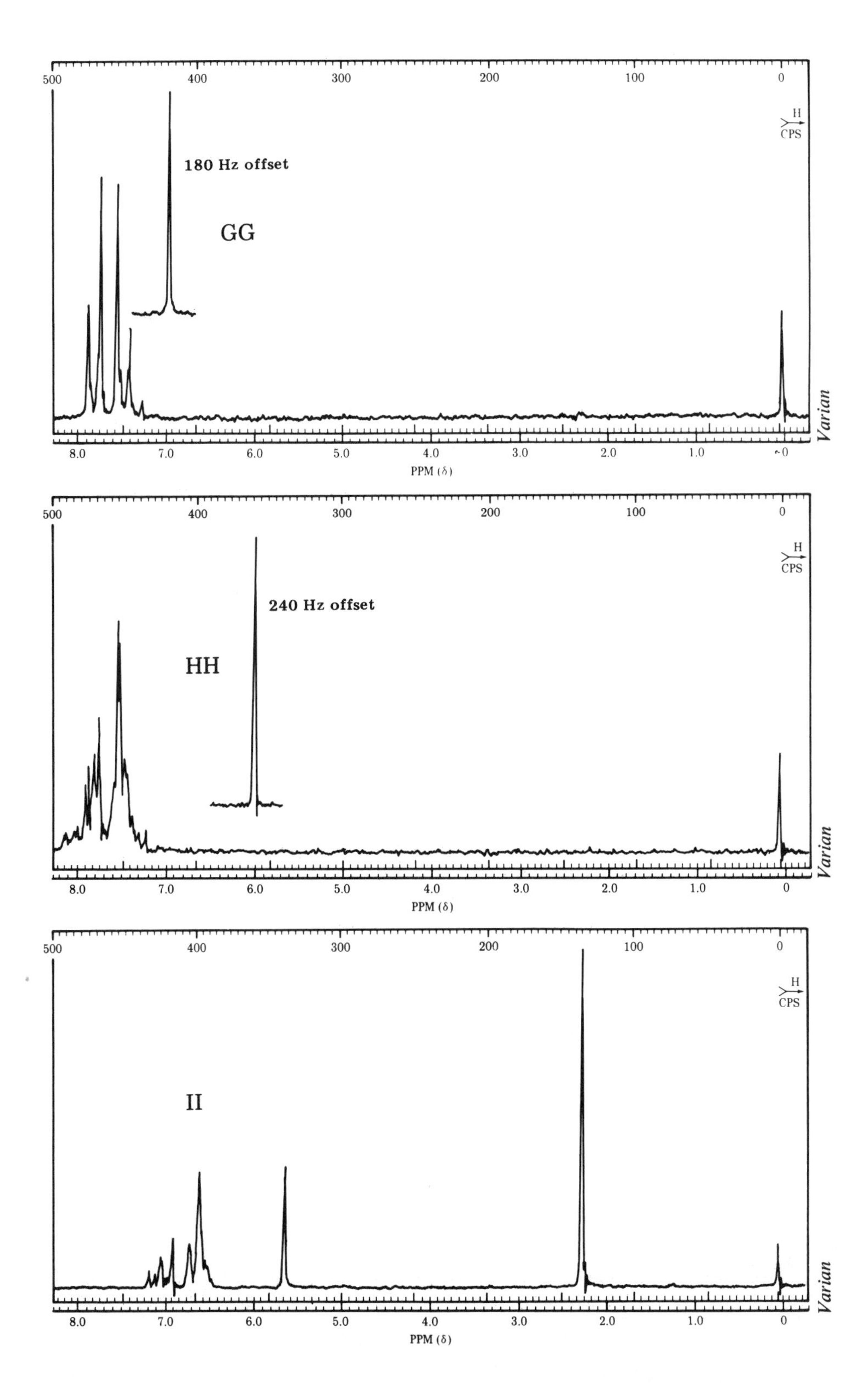

500
400
300
200
100
0
H
CPS
180 Hz offset
GG
Varian
8.0
7.0
6.0
5.0
4.0
3.0
2.0
1.0
0
PPM (δ)
240 Hz offset
HH
II

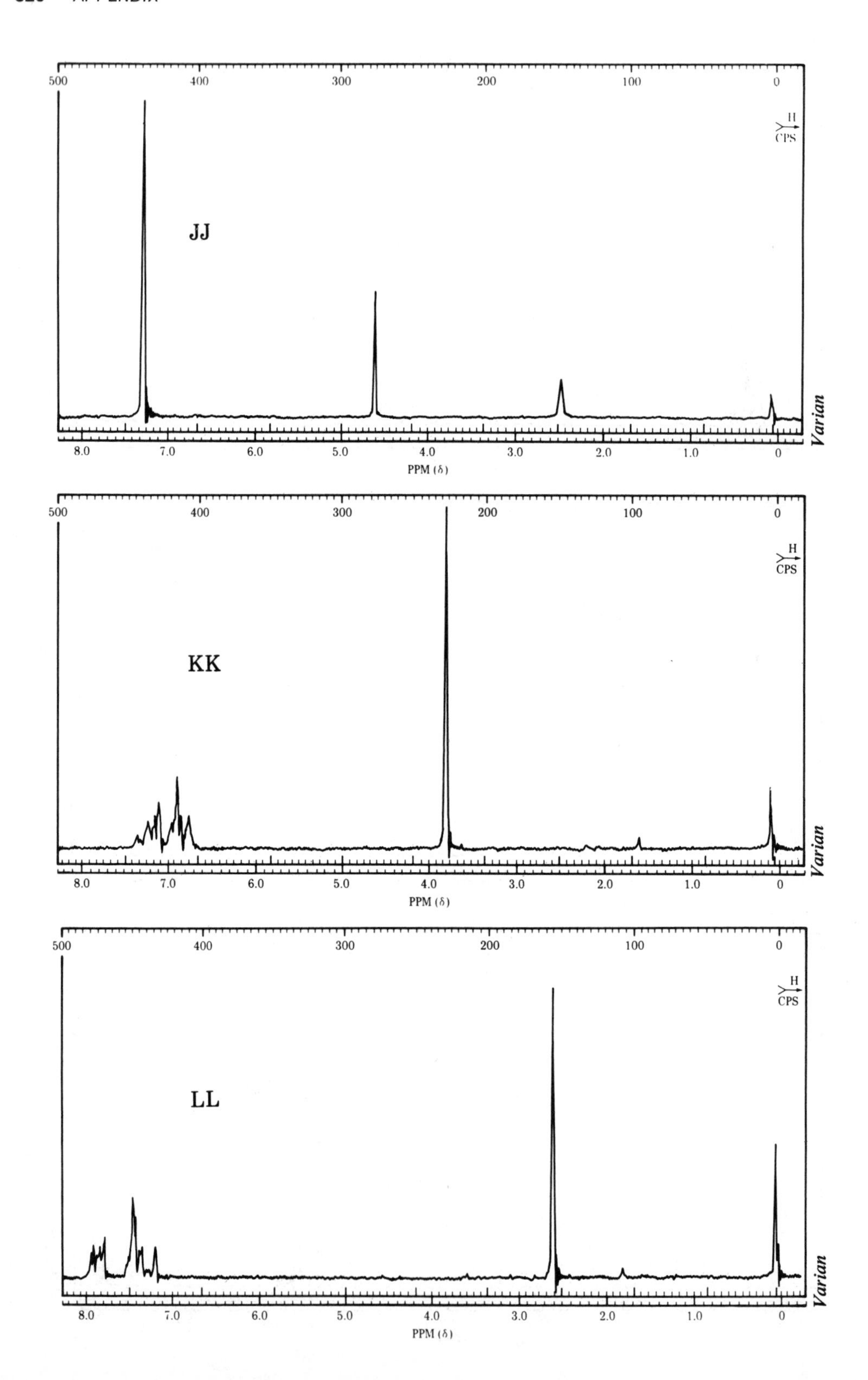

JJ
KK
LL
500
400
300
200
100
0
H
CPS
8.0
7.0
6.0
5.0
4.0
3.0
2.0
1.0
PPM (δ)
Varian

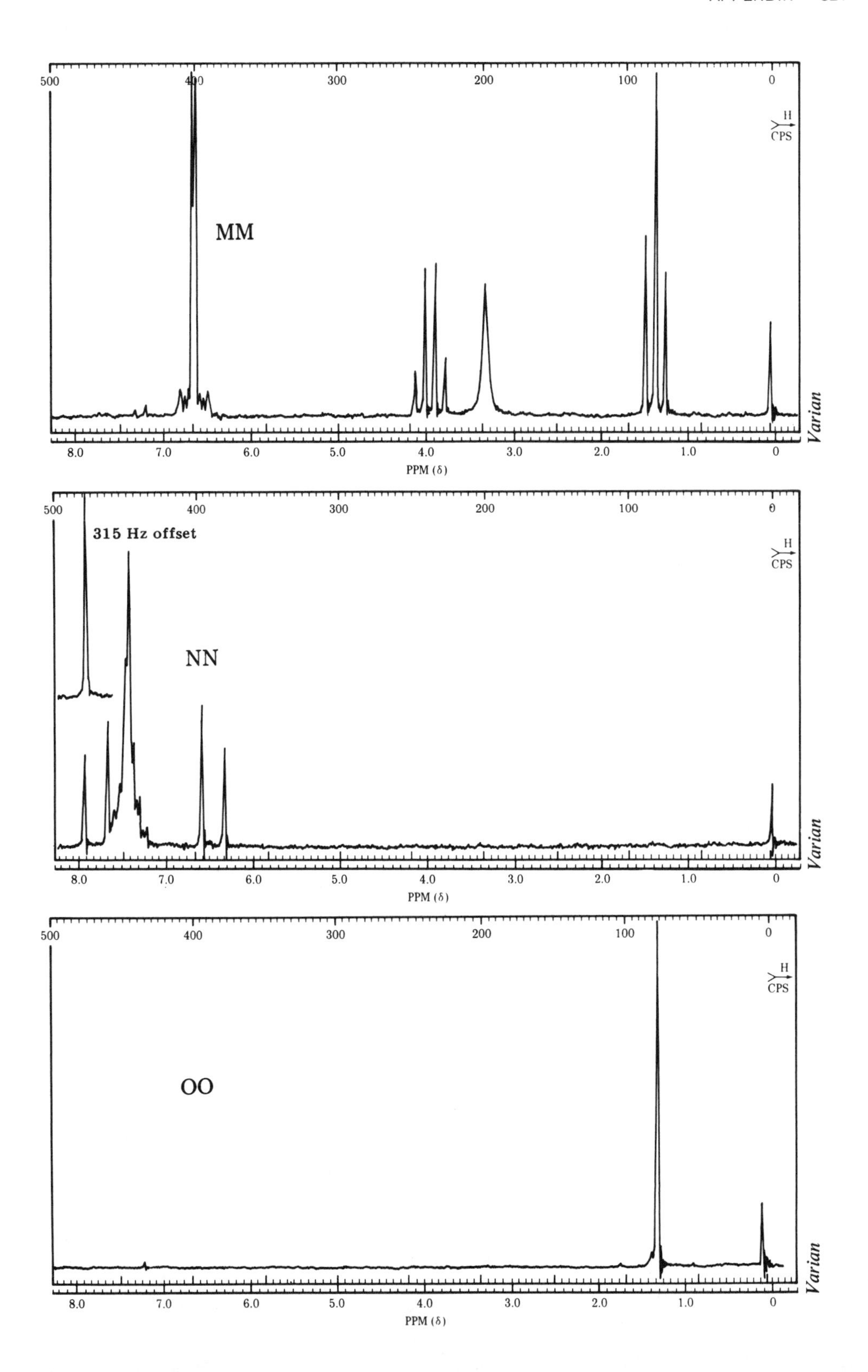

500
400
300
200
100
0
H
CPS
MM
8.0
7.0
6.0
5.0
4.0
3.0
2.0
1.0
0
PPM (δ)
Varian
315 Hz offset
NN
OO

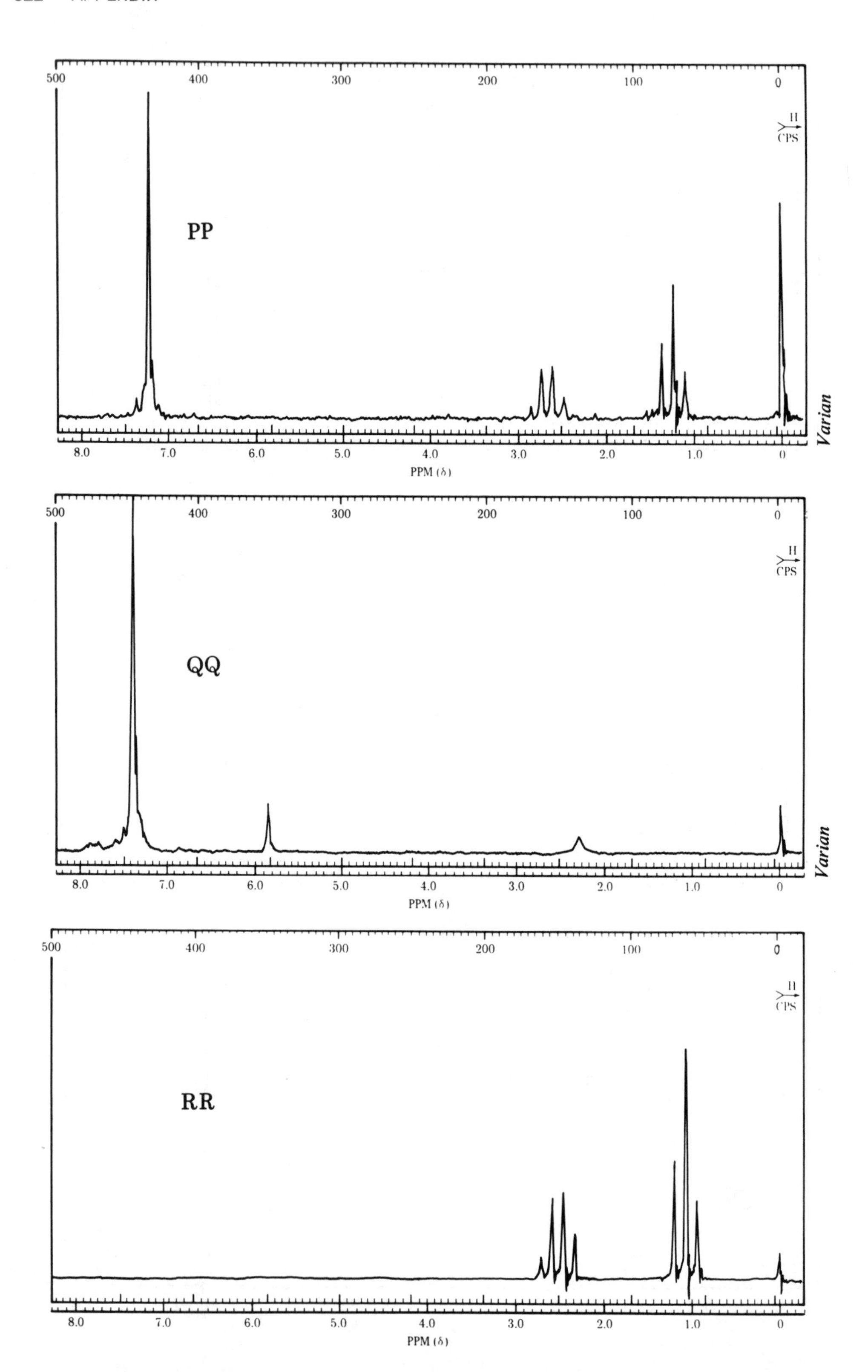
PP
QQ
RR
500
400
300
200
100
0
H
CPS
8.0
7.0
6.0
5.0
4.0
3.0
2.0
1.0
0
PPM (δ)
Varian

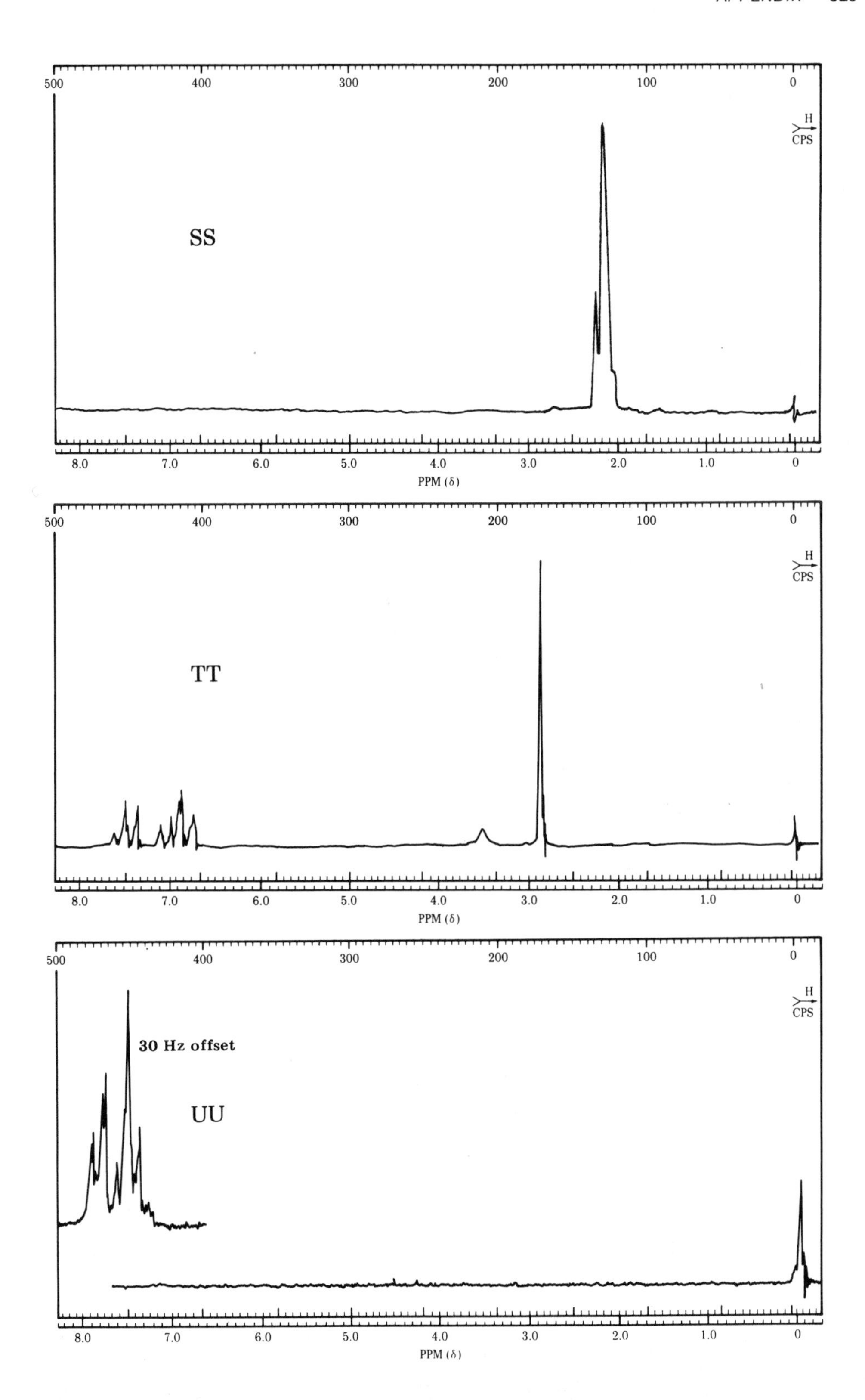
500
400
300
200
100
0
H
CPS
SS
8.0
7.0
6.0
5.0
4.0
3.0
2.0
1.0
0
PPM (δ)
TT
30 Hz offset
UU

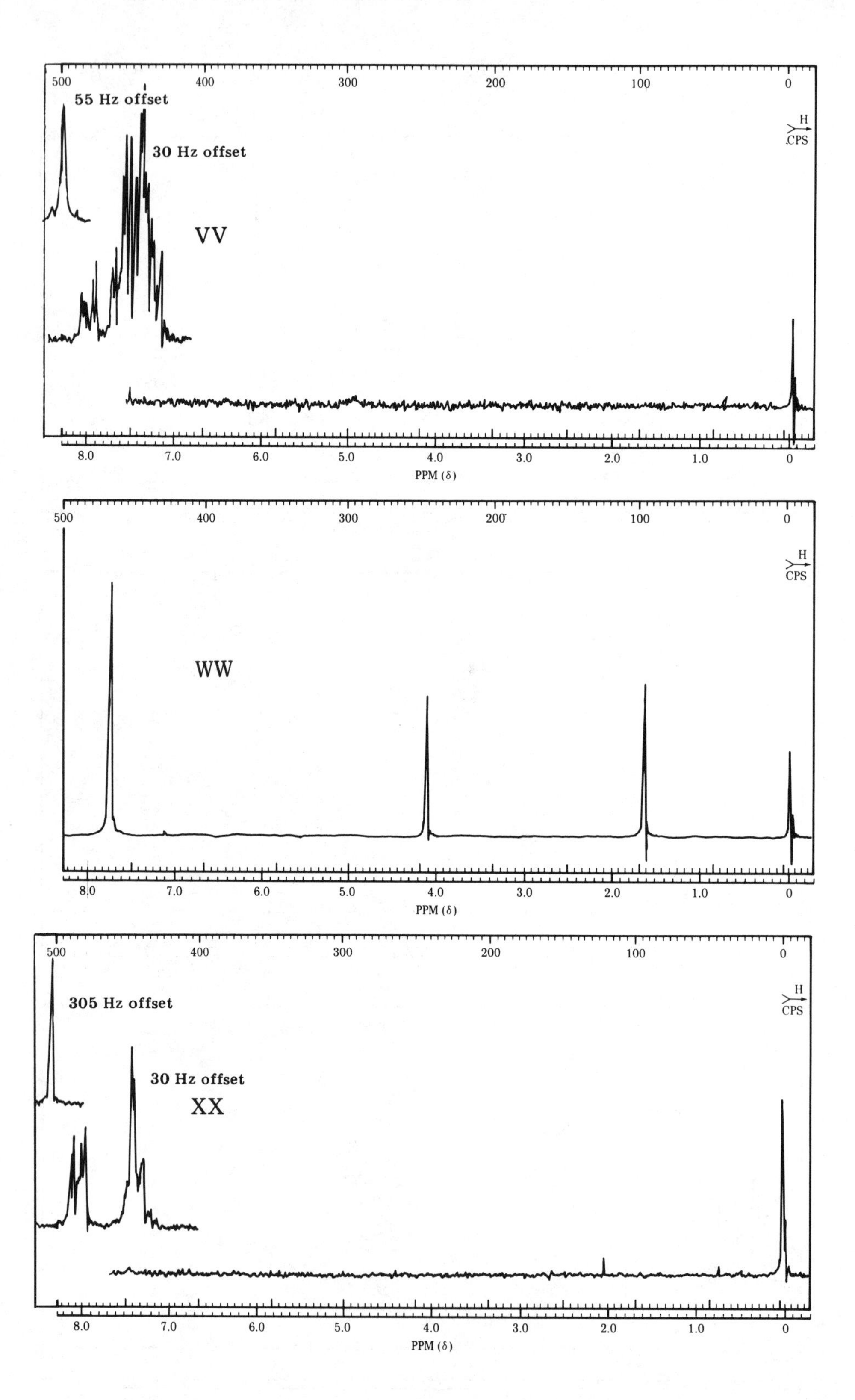

500
400
300
200
100
0
55 Hz offset
30 Hz offset
VV
H
CPS
8.0
7.0
6.0
5.0
4.0
3.0
2.0
1.0
0
PPM (δ)
WW
305 Hz offset
30 Hz offset
XX

3 4 5 6 7 8 9 0